SOUND COLOR

Library of
Davidson College

SOUND COLOR

Wayne Slawson

UNIVERSITY OF CALIFORNIA PRESS
BERKELEY   LOS ANGELES   LONDON

UNIVERSITY OF CALIFORNIA PRESS
BERKELEY AND LOS ANGELES, CALIFORNIA

UNIVERSITY OF CALIFORNIA PRESS, LTD.
LONDON, ENGLAND

© 1985 BY THE REGENTS OF
THE UNIVERSITY OF CALIFORNIA

Library of Congress Cataloging in Publication Data

Slawson, Wayne.
Sound color.

Bibliography: p.
Includes index.
1. Tone color (Music) 2. Music—Acoustics and physics.
3. Sound. 4. Music—Psychology.
5. Composition (Music) I. Title.
ML3807.S53   1985   781'.15   84-2474
ISBN 0-520-05185-8

PRINTED IN THE UNITED STATES OF AMERICA
1   2   3   4   5   6   7   8   9

*To Madeleine Slawson
and in memory of
Wilbert W. Slawson
with love and respect*

Contents

Preface   xv

Acknowledgments   xxi

Table of Recorded Examples   xxiii

CHAPTER ONE: Precursors and Premises   3

    Previous Studies of Timbre or Sound Color   5

        Pierre Schaeffer's Traité des objets musicaux *(1968)*   5,
        Robert Erickson's Sound Structure in Music *(1975)*   10,
        Robert Cogan and Pozzi Escot's Sonic Design *(1976)*   11

    Formal Requirements of Sound-Color Theory   14

        Elementary Units   15,   *Psychoacoustic Definition of Elements*   15,   *Invariance versus Difference*   16,   *Operations on Sound Color*   17

    Timbre versus Sound Color: Limiting the Scope   18

    The Principal Questions   20

CHAPTER TWO : A Theory of Sound Color   22

    The Source/Filter Model: Three Examples   22

        The Wall Tapper   22,   *Vowels in the Human Vocal Tract*   24,   *Musical Instruments and "Strong Coupling"*   27,   *A Cautionary Note*   31

    Resonances and the Spectrum Envelope   31

        Simple Resonance: The First "Hill"   32,   *Multiple Resonators: Uniform Tubes*   35,   *Resonances in a Uniform Tube*   35,   *How to "Add" Resonance Curves*   36,   *Formants*   38,   *Spectrum Envelopes of Non-Uniform Tubes*   39

            F-Patterns   39,   F-Patterns and Spectrum Envelopes   40

        Sound Color and Resonances: Rule 1   41,   *Limitations in the Scope of Rule 1*   42,   *The Adequate Source*   43,   *Ideal Sources*   44

            The Sinusoidal Source   44,   The Impulse   44,   The Pulse Train   45,   Noise   46

        Guidelines for Adequacy in Sources   46,   *Distinguishing Source from Filter*   46

    The Dimensions of Sound Color   48

        Distinctive Features and Sound-Color Dimensions   51,   *Equal-Value Contours: Specification of the Dimensions*   54

            OPENNESS   54,   ACUTENESS   55,   LAXNESS   56,   SMALLNESS, an Additional Dimension   56

        Sound-Color Space: Rule 2   57,   *Measurement in Sound-Color Space*   58,   *Independent Constancy versus Independent Variation*   59,   *Alternative Formulations of the Dimensions*   60,   *The Sound-Color Dimensions as Pre-Speech*   60,   *Normalization of the Sound-Color Space*   62,   *Limitations of the Dimensions*   65

Resolution of F-Pattern Analysis 65, The Domain of Color-Space Shifts 66, Sound Color in Aberrant Resonators 66

Operations on Sound Color 68

General Characteristics of Operations 68, *Sound-Color Transpositions* 69

LAXNESS Transposition 71, Wrap-Around in Color Transposition 72, Wrap-Around in SMALLNESS Transposition 73, LAXNESS Transposition with Wrap-Around 75

The Transposition Rule: Rule 3a 75, *Sound-Color Inversions: Rule 3b* 76

ACUTENESS Inversion 77, OPENNESS Inversion 78, SMALLNESS Inversion 78, Effects of Inversions on LAXNESS 78

Operations in "Shifted" Spaces 79, *The Rationale: Are Color Transposition and Inversion Real?* 79

The Artificiality of the Sound-Color Operations 81, Why Is the Neutral Color the Axis of Inversion? 81

Combinations of Operations 84, *Extensions of the Operations* 84

Color Hierarchies 85, Hierarchies by Means of Successive Filtering 85, Hierarchies by Sound-Color Space Limitations 86

Color Dynamics 87, *Color Mixture* 88, *Possible Additional Dimensions* 89, *Summary of the Sound-Color Rules* 89

CHAPTER THREE: Evidence from Auditory Physiology 91

Anatomy of the Auditory System 93

The Cochlea 93, *The Pathways of the Auditory Nervous System* 94

Frequency Analysis in the Periphery: Classical Views 97

Resonance Analysis 100

Vowels and the Auditory Nerve 101, *Resolution of Resonances* 104, *Possible Cochlear-Nucleus Responses to Resonances* 105, *Resonance Detectors in the Periphery?* 105

F-Pattern Detectors 106

Differentiation of Source from Filter 108

The Physiological Correlates of Pitch 109, *Periodicity Detectors* 110

Dimensions and Operations 112

Open Questions: A Model and Some Proposed Physiological Research 113

CHAPTER FOUR: Evidence from Psychoacoustics 116

Limitations of Psychoacoustic Methods 116

Threshold Measurements 117, *Measures of Psychological Distance* 118, *The Problem of Context* 119, *The Problem of Learning* 119

Psychoacoustics of Filter Systems 120

Sensitivity to Changes in Filter Parameters 120, *Interpretation of the Sensitivity Measures* 122, *Mechanisms for Resonance Detection* 123, *Psychoacoustic Scales for Resonance Frequency* 125

Psychoacoustics of Source Characteristics 127

Registral Pitch: The Mel Scale 127, *Periodicity or Interval Pitch* 129

Psychoacoustic Comparisons of Source and Filter Characteristics 130

Cues for Normalization 133

Psychoacoustics of the Sound-Color Dimensions 134

Dimensions of Complex Sounds 134, *Multidimensional Scaling Studies* 134, *Dimensions of Musical-Instrument Timbre* 138

Psychoacoustics and the Sound-Color Operations 140

Conclusions and Open Questions 141

CHAPTER FIVE : Evidence from Speech and Cognitive Science 144

Limitations: "Active" Theories of Speech Recognition 145

Evidence Concerning the Speech Mode 146

Categorical Perception 147, Hemispheric Specialization: Clinical Results 148, Hemispheric Specialization: Dichotic Listening 149, Perception of Speech in Animals 151

Perception of Vowels and Sound Colors 152

Vowel Perception in Man 153, *Normalization of Vowels* 155, *Temporal Aspects of Vowel Perception* 156, *Vowel-like Resonances in Some Musical Instruments* 157

Dimensions of Sound Color: Multidimensional Scaling of Speech and Non-Speech 158

Sound Color Operations: Analogies from Cognitive Psychology 159

Symmetry about the Vertical versus Other Transformations 161, *Mental Rotation* 162

Conclusions and Open Questions 163

CHAPTER SIX : Musical Evidence: Sound Color in Electronic Music 165

Limitations of Music-Analytic Evidence 166

Independent Musical Control of Sound Color 168

Color as a Contrapuntal Differentiator 168, *Invariant Sound Color as a Structural Link* 170, *Color in the Foreground* 173

Evidence for the Dimensions 175

Change along Specific Dimensions 176, *Exposition of the Sound-Color Space* 177, *Completion Based on Sound Color* 180

Operations on Sound Color? 183

A Special Case: Stockhausen's *Kontakte* 184

Vocal Music and the Sound Color Theory 184

Timbre Pieces Organized in Alternative Ways 186

Summary and Conclusions 189

CHAPTER SEVEN: Composition with Sound Color 190

Closure, Normality, and Cyclic Notation 192

Properties of Some Normal Sound-Color Collections 196

Twelve-Color Collections 196, *Collections with LAXNESS Closure* 197, *Compositional Implications of Closure and Non-Closure* 198

Multiple Operations on the Normal, Nine-Element Collection 199

A Convention for Naming the Operations 200, *Rules for Combining Operations* 201, *The Unique Inversions on the Normal Nine-Color Collection* 203, *The Unique Transpositions on the Normal Nine-Color Collection* 204, *Combinations of Transpositions and Inversions* 205, *Subsets of the Normal Eight- and Nine-Color Collections* 206

The Composition *Colors* 207

The Overall Form of Colors 208

The Basic Sound Color Materials 209, *Properties of the Ordered Series of* Colors 210

    Operations on the Series 210, Color Combinatoriality 213, Combinatorial Sound-Color Structure in *Colors* 214

Setting the Combinatorial Color Structure 216

    The Settings in Variations 1–3 217, The Remaining Variations: A Summary 220

Conclusion and Perspectives 222

CHAPTER EIGHT: A Critique and Afterword 224

APPENDIX: Using Filters to Control Sound Color 226

    Filters in the Analog Electronic Music Studio 226

    Filters in Computer Music 228

Bibliography 231

Index 250

Preface

AMONG THE IMPORTANT musical developments of the twentieth century is an increasing interest in elements of music other than pitch and rhythm. In particular, timbre—controlled, in performance, by innovative techniques of playing and elaborate instrumental combinations and, in electronic and computer music, by direct manipulation of electronic signals—has taken on a prominence that is in marked contrast to its role in earlier music.

This book is about an aspect of timbre. Its main purpose is to provide some guidance to composers exploring that fascinating aspect of sound. The book is a contribution, more or less, to music theory. It is less so than most music theories because its main ideas are few in number, relatively undeveloped, and inexact in form, and because it deals more with musical potential than with music already composed. It is more so because it delves rather deeply into the empirical basis of the theoretical ideas and because it is largely concerned with what Robert Morris calls "compositional design": the process of planning and working out that has been an important part of composition in the European tradition and elsewhere for centuries.

This book is *not* about instrumentation or orchestration. The acoustical properties of some musical instruments are discussed in passing, but the main thrust of the discussion is toward an abstract conception of timbre as an aspect of auditory sensation. It may be that some of the principles of the theory can be applied in instrumental music, but that will have to await some future project in research and/or composition. The theory can be applied directly, however, in electronic and computer music and in music for voice and vocal ensembles.

The book is addressed first of all to musicians and musical scholars, and particularly to composers of electronic and computer music and music for the voice and to music theorists. The main motivating forces behind the book were a felt need for a means of organizing an aspect of my own music, and my hope that other composers and theorists will find in the theory of sound color a basis for furthering their work. I also hope to attract a secondary audience of scientists, especially sensory and cognitive psychologists and auditory physiologists. The theory of sound color presented here is both a psychological and a musical theory, and I hope it may suggest experimental studies in those and appropriate other scientific fields.

I have tried to keep unnecessary formalisms out of the book at the expense of sometimes lengthy explanations and, particularly in the scientific chapters (Three, Four, and Five), I have attempted to introduce key concepts at an elementary level. Nevertheless, readers with particular backgrounds will probably find parts of the book more interesting than others. Musicians, no doubt, will want to concentrate on Chapters One, Two, Six, and Seven. Psychologists and other scientists will probably find the first five chapters of most use to them. I hope that all readers will try at least to skim the portions of the book that are outside their fields of specialization. No single source of evidence is sufficient to lend very great credence to the theory; taken as a whole, however, the scientific and musical evidence presents a much stronger case. Only by bringing to bear all the evidence can a fair judgment of the theory be made.

I have used a draft of the book as the main text for a graduate course on timbre for composers. The class, made up of academically well-prepared and curious students, seemed to take to it fairly well. At least portions of the book could be used, then, as a supplementary text for a graduate-level survey of music theory and for graduate and upper-level undergraduate courses in electronic and computer music, music acoustics, and sensory psychology. I have assumed some acquaintance with Fourier analysis, some elementary knowledge of musical acoustics, in Chapter Three a smattering of high school biology, and in Chapters Four and Five a little psychology. Chapters 4 and 5 of Lindsey and Norman (1977) or Slawson (1975) should suffice to provide the needed background in all these areas.

In the development of part of the theory, I have found it necessary to refer to the sounds of certain vowels. Instead of using the International Phonetic Alphabet to represent those sounds, I decided to adopt a two-letter convention that I believe to be more evocative of most English speakers' phonetic intuitions. Since the sounds are not literally vowels but are, rather, generalized forms of vowels, a convention apart from the IPA seemed doubly justified. In the accompanying table is a list of the correspondences among my convention, the IPA, and some English (and German) words containing the vowels.

SOUND COLOR	IPA SYMBOL	PRONUNCIATION
uu	u	b*oo*t
oo	o	b*oa*t
aw	ɔ	b*ou*ght
oe	ø	b*ö*se (German)
aa	a	h*o*t
ii	i	b*ee*t
ee	e	b*ai*t
ae	æ	b*a*t
uh	U	p*u*t
ah	ʌ	b*u*t
ih	I	b*i*t
eh	ɛ	b*e*t
ne (neutral)	ɜ	th*e* (unstressed)

The history of this study is long and checkered. In retrospect I can see that the germs of the ideas in the book were planted during informal studies of the acoustics of speech with Gordon Peterson in 1958. Those studies were undertaken in an attempt to find ways of applying knowledge about speech sounds to music composition. Only twenty years later was I able to see what seems a coherent and musically natural way of reaching that goal. Along the way I had a variety of different experiences that—again, in retrospect—contributed bits and pieces of a solution to the puzzle. The first ingredient was my musical development under the guidance of the teachers who most influenced me, Ross Lee Finney, Leslie Bassett and Hans David, and later of my friends and former colleagues Robert Morris and Allen Forte. Study of speech production and perception with Kenneth Stevens at MIT and Gunnar Fant in Stockholm was another ingredient. Still another was my introduction to sensory and cognitive psychology with S. S. Stevens, George Miller, and Eric Lenneberg at Harvard. Finally, my work with computer-synthesized sound, principally at the Mitre Corporation and in Marvin Minsky's laboratory at MIT, contributed in an important way.

Two purposes are served by indulging in this bit of nostalgia. The first is to acknowledge the indirect, but no less critical, help that I have received from all those distinguished musicians and scholars. I am sincerely grateful for that help and I am sorry indeed that three of them have died before I have been able to complete this book, a concrete result of their tutelage.

Second, I hope this personal history of dilettantism will explain—perhaps even excuse—the quite disparate points of view that are brought together in this book. I can only plead that the goal of my study, establishment of a musically viable means of dealing with a new musical element, requires such a many-faceted approach. The price to be paid is an unusually heavy demand on almost all readers to wade through some portion of the book that is outside their field of expertise.

I have had much direct help from a number of people during the research for and the writing of this book. Many discussions with Robert Morris, and his careful reading of two drafts, have greatly improved the book. An early criticism of his led me to alter the theory in a substantive way. Earl Schubert provided valuable support at an early stage and perspicacious criticism of an early draft, particularly of Chapters Three, Four, and Five. Catharine Slawson is responsible for a marked improvement in the style and flow of Chapter One. John Spitzer edited the entire manuscript and his clear-minded skepticism aided me immeasurably in eliminating a good many murky passages. Eugene Deskins, C. Y. Chao, and John Peel helped to clarify my thinking about certain group-theoretical points. Phillip Rubin, of Haskins Laboratory, helped me with some of the research reported in Chapter Six. Campbell Searle caught several mistakes in Chapter Two. Students in my class, Tim Buell, Cheryl Bruner, Gil DeBenedetti, Jim Lovendusky, Chinny Ohia, and Reza Vali, were the experimental subjects, whose responses in the form of comments and questions pointed out a number of gaps in the presentation, awkward explanations, and infelicities of style. Two anonymous readers made several valuable suggestions that I was able to incorporate in the book. Alain Hénon shepherded the book through the editorial process. Lucky indeed is the author whose work is copy-edited by Jane-Ellen Long. She detected several substantive errors and faults of organization. Her knowledge, sense of style, tact, and sympathy with the author's sometimes unexpressed purpose were truly extraordinary. It is a pleasure to thank all the people mentioned here, and the many others who have contributed in smaller ways.

Three institutions gave me significant support at critical junctures: A sabbatical leave from the University of Pittsburgh and a Fellowship from the American Council of Learned Societies enabled me to spend full time for nearly a year on the study. Stanford University, in particular the Center for Computer Research in Music and Acoustics headed by John Chowning, provided a hospitable home-base and a fine computer editor program during that year. I am most grateful for that quite unusual help and encouragement.

Acknowledgments

I WISH TO ACKNOWLEDGE with thanks the permission granted by the following authors and publishers to use figures from their publications:

Figure 1: Éditions du Seuil.
Figure 10: Mouton Publishers, a division of Walter de Gruyter & Co.
Figure 22: M. M. Merzenich and Academic Press.
Figure 23: I. C. Whitfield.
Figure 24: M. B. Sachs and the American Physical Society.
Figure 25: W. D. Keidel and Springer-Verlag.
Figure 26: G. Langner and Plenum Publishing Corporation.
Figure 28: The American Physical Society.
Figure 29: The American Physical Society.
Figure 30: R. Plomp.
Figure 31: G. von Bismarck and S. Hirzel Verlag.
Figure 32: The American Physical Society.
Figure 33: Gunnar Fant.
Figure 36: Universal Edition A. G.

Recorded Examples

THE RECORDS that accompany this book contain examples illustrating certain aspects of the theory of sound color and portions of the author's composition *Colors*. The page numbers indicate where each recorded example is discussed in the text.

SIDE ONE

1A. Pitch change in weakly coupled systems (p. 29). Three sounds with different filter settings are excited with a buzz source at a fundamental frequency of 65 Hz and then again with a similar source at 130 Hz.

1B. Pitch change in strongly coupled systems (p. 29). The first three sounds are the same as in 1a; in the second three sounds, both the source and the filter settings are doubled in frequency.

2A. Three filter settings excited by impulses (p. 44).

2B. Three filter settings excited by pulse trains having different repetition rates (p. 45).

2C. Three filter settings excited by noise (p. 46).

3A. A series of three sound colors transposed in the direction of increased ACUTENESS (p. 70).

3B. The series of 3a transposed in the direction of increased OPENNESS (p. 70).

3C. The series of 3a transposed in the direction of increased SMALLNESS (p. 70).

4. LAXNESS transposition (p. 75). Three sound colors (excited by noise) are repeated five times, gradually becoming increasingly LAX, until all three are the neutral sound color, then the original sound colors are restored in a sixth repetition. The entire example is repeated with a pitched source.

5A. A series of four sound colors is inverted with respect to ACUTENESS (p. 77).

5B. The series of 5a inverted with respect to OPENNESS (p. 78).

5C. The series of 5a inverted with respect to SMALLNESS (p. 78).

SIDE TWO

6A. Evidence for the LAXNESS dimension (p. 161). Four sound colors are repeated with gradually increasing LAXNESS until the maximally LAX position is reached; then LAXNESS is gradually decreased until the original four colors are heard again.

6B. Example 6a is repeated with a pitched source and rhythmic pattern that emphasize the groups of four colors (p. 161).

7A. Excerpt from the beginning of Milton Babbitt's *Ensembles for Synthesizer* (p. 178).

7B. Samples from the four "static" sounds from Example 7a (p. 178).

8A. Sound color combinatoriality (p. 213). The prime series of nine colors is excited by amplitude-modulated noise; the retrograde of the OPENNESS inversion, by a pitched source.

8B. Another combinatorial matrix (p. 213). The prime series is excited by a pitched source; the retrograde of the OPENNESS inversion, by amplitude-modulated noise.

SIDE THREE

9. Variation 1 from *Colors* (p. 217).

10. Variation 3 from *Colors*, first section (p. 220).

SIDE FOUR

10. (continued). Variation 3 from *Colors*, concluded.

11. Variation 11 from *Colors* (p. 221).

SOUND COLOR

CHAPTER ONE

Precursors and Premises

The evaluation of tone color, the second dimension of tone, is in a much less cultivated, much less organized state than is the aesthetic evaluation of [pitch]. Nevertheless, we go right on boldly connecting sounds with one another, contrasting them with one another, simply by feeling; and it has never yet occurred to anyone to require of a theory that it should determine laws by which one may do that sort of thing. . . . Now, if it is possible to create patterns out [of] pitch[es], patterns we call "melodies," progressions, whose coherence evokes an effect analogous to thought processes, then it must also be possible to make progressions out of . . . "tone color," progressions whose relations with one another work with a kind of logic entirely equivalent to that logic which satisfies us in the melody of pitches. ARNOLD SCHOENBERG ([1911] 1978)

SCHOENBERG WROTE ALMOST nothing about "tone color melody" after these famous remarks that close his *Harmonielehre*.[1] The equally famous "Farben" movement of the *Five Pieces for Orchestra* (Schoenberg 1912), composed only two years before the book was first published, is a beautiful and radically novel experiment with tone-color structure. But Schoenberg wrote no other music of that kind. Throughout his works we find fascinating scoring, scoring that sometimes takes on structural significance, but nowhere, after "Farben," is orchestration the most prominent musical element. Did Schoenberg lose interest in "tone color melody"? Did he decide that the "Farben" experiment was not successful? Did he find it impossible to construct the theory he speaks of in the *Harmonielehre* and, if so, was the lack of a theory a block to further compositional developments? We cannot know Schoenberg's reasons for turning to other matters, but we are led

---

[1] In *Style and Idea*, the composer's selected writings, we have only a short note—written in 1951, forty years later—defending his invention of the term in the face of a claim of priority attributed to Webern (Schoenberg 1975).

to suspect that the study of timbre[2] must present extraordinary difficulties, for only a few others have dealt with it seriously and at length, and no composer has continued along the lines begun in "Farben." Very little music in which timbre is the primary element has appeared in the seven decades since Schoenberg's challenging statement was written and his provocative piece composed.

Motivated by the assumption that composition with sound color requires a theory, in the present study I attempt to take up where Schoenberg left off. My central purpose here is to present and defend a set of hypotheses that can serve as "laws" for the combination of what is called *sound color*. This study does not deal directly with the "color" of musical instruments, however. Rather, it is an investigation of color in a more general sense, as a *dimension* of sound.[3] The choice of concentrating on this abstract conception of timbre follows from two considerations. First, the properties of musical instruments are such as to make construction of a general theory of orchestration or instrumen-

---

[2] Many terms for this aspect of sound—*tone color, sound quality, sound color, instrumental color*, in addition to *timbre*—have been used in a variety of contexts. Although each of the terms sometimes appears to have its own shades of meaning, there is little consistency and the terms are often treated as synonymous. One goal of the present study is to establish a more precise terminology (see discussion below, page 18).

[3] The "Farben" movement is somewhat misleading in this connection. It was organized, as Burkhart (1973) has shown, according to an intricate and regular permutation of instrumental combinations. Whether Schoenberg intended that we should be able to follow the structure of the piece by listening to the "dimension" of the resultant tone color of those fleeting instrumental combinations or whether he intended that we identify the instruments in each combination in order to hear the structure, we have no way of knowing. If the remarks in *Harmonielehre* can be taken as a clue, we would suspect that he had in mind the former, more abstract mode of listening. That mode is also suggested by the restricted ranges of the pitches and dynamics, which favor tone-color blending. The piece is organized according to the second mode of listening, however. These two modes would be the same only if the tone color, in the sense of a dimension of sound, were the same when the instrumental combinations are the same—a condition that seems on the face of it to be satisfied only rarely. This confounding of tone color with instrumental combinations may have presented difficulties to Schoenberg's ear that led him to abandon further developments along the same line.

tation immensely complex and difficult. Second, it is now possible to control sound in such a way as to experiment with timbre precisely, without taking into account the motor abilities of human performers or the properties of wood, metal, and strings, as musical-instrument designers must do. In electronic and computer-music studios, the musical by-product of a revolution in electronics that took place in the seventy years since the publication of *Harmonielehre*, the acoustical foundations of timbre can be studied essentially without limitation and those studies applied to music composition. This book is about sounds produced by electronic means and (incidentally) by the human voice.

PREVIOUS STUDIES OF TIMBRE OR SOUND COLOR

No corpus of "laws" of the sort Schoenberg called for—none, in any case, that provides a means of organizing sound color according to its own musical logic—exists at the present time. However, three important studies on timbre or sound color have appeared since Schoenberg's book: Pierre Schaeffer's *Traité des objets musicaux*, Robert Erickson's *Sound Structure in Music*, and Robert Cogan and Pozzi Escot's *Sonic Design*. Each of these books has significantly influenced composers and musical scholars, and each is an important precursor to the present study.[4]

Pierre Schaeffer's *Traité des objets musicaux* (1968)

Regrettably, Schaeffer's monumental study is not well known in the United States. No prominent American music journal has given it an extensive review.[5] The book's approach to composition and musical study has been influential, however, in France and elsewhere in Europe—notably, in Utrecht and Stockholm. Much of *Traité* is a philosophical defense of an attitude toward sonic experience that derives from the *musique concrète* tradition. In the core of the study, Schaef-

---

[4] A brief article by James Tenney (1965) is insightful, but essentially encompassed by Schaeffer's book.

[5] For brief reviews in English, see Kay (1968) and Evangelisti (1967).

fer describes an ambitious research program whose aim is nothing less than the classification of all sound, and a pedagogy—a *solfège*—to train musicians in the musical use of the classification scheme. The training and research programs have been implemented at the French National Radio in Paris (GRM) and they are presently being actively pursued by a group of devoted researchers and students.

Essential to Schaeffer's approach is the development of the concept of *reduced hearing*. The common mode of listening, in which we respond to a sound by identifying its source—the sound "is" an oboe, a jet plane, a whippoorwill—must be distinguished, according to Schaeffer, from another mode, in which we purposely—perhaps in some sense automatically—divorce what we hear from its source, concentrating instead on the properties of the sound itself. This kind of objectification or "reduction" of sound is required for a sonic event to be heard as a "sound object" (*objet sonore*).[6] Since sounds made by different things will often sound alike when "reduced," the number of sound objects is smaller than the universe of all possible sounds. These sound objects, not all things that make sounds, are the subjects of Schaeffer's first grand classification.

Sound objects are judged according to two criteria, *mass* (*masse*) and *treatment* (*facture*). The five categories of mass are: M_1, pure tones; M_2, complex pitched sounds; M_3, complex, non-variable sounds without pitch (*masse fixe*); M_4, sounds that vary somewhat; and M_5, sounds that vary considerably. Schaeffer uses *mass* to classify sounds according to the kinds of frequency spectra they exhibit. *Treatment* is more complicated. It describes "the way in which the sound is communicated or made manifest throughout its duration" (p. 432). Seven categories of treatment are divided into three groups, the first group consisting of three categories (F_1, F_2, and F_3) of "continuous" sounds; the second group, a single category (F_4) of impulsive sounds; and the third group, three categories (F_5, F_6, and F_7) of "discontinuous" sounds that mirror those in the first group (see Figure 1).

---

[6] Reduced hearing seems a prerequisite for the abstract mode of listening that hears timbre as a dimension of sound, and is apparently referred to by Schoenberg (quoted above).

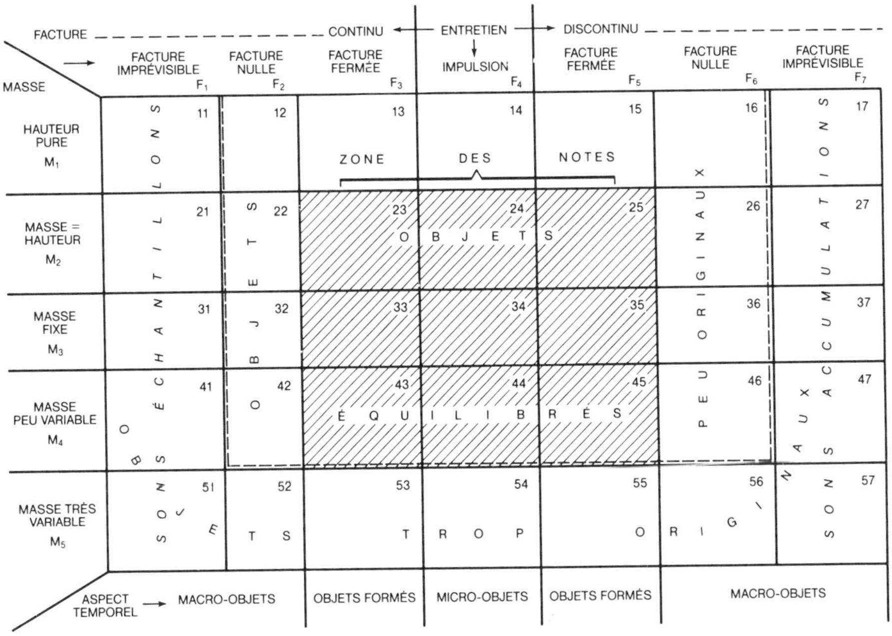

FIGURE 1: *Schaeffer's scheme for classification of sound objects (from* Traité des Objets Musicaux, *p. 442).*

The categories in the scheme can be illustrated with a few examples. A sine wave lasting for an indefinitely long duration would be placed in the first categories of both mass and treatment. Schaeffer calls the sounds in the first treatment category "sampled sounds"; the classifier considers only a brief portion of the total, indefinitely long duration. A note played by an oboe on a fixed pitch with a duration of, say, half a second would be placed in the second category of mass and the third category of treatment. A slow roll on a cymbal with hard sticks is unpitched and yet discontinuous, so it would fall into categories M_3 and F_5. Schaeffer cites, as examples of the "eccentric" sounds of the last category of treatment and the "highly variable" category of mass, passages of Xenakis's orchestral music that are made up of "clouds" of pizzicato and glissando strings.

Schaeffer's treatise culminates with a discussion of what he calls the *musical object* (*objet musical*). Here Schaeffer is not entirely clear. Some passages give the impression that musical objects are to be construed as making up a subset of the totality of sound objects—those that have musical potential essentially as a matter of definition. Elsewhere the question of context is brought in; a sound object (any sound object?) placed in a suitable musical structure becomes a musical object. It may be that Schaeffer intends the question of what is or is not a musical object to be left somewhat open; if, in other words, a suitable musical context for a sound object can be found, now or in the future, then it becomes a musical object.

Schaeffer proposes a second classification scheme for musical objects. As one would expect when questions regarding the musicality or the musical potential of a sound are raised, many more categories are required than in the case of the sound object. Schaeffer speaks of the *type* of a sound, its *class*, its *genus*, and its six different *species*. In addition, seven "criteria" are described: *mass* (from the sound-object analysis), *dynamic* (as opposed to static), *harmonic timbre, melodic profile, mass profile, grain,* and *allure*. A musical object does not simply belong to one of the categories, nor is it to be judged according to only one of the criteria. Rather, the categories and criteria provide a systematic program for evaluation of musical sounds.

The researcher or student first considers a sound according to each of the criteria, then the sound's "musical morphology" or class, its "musical character" or genus, and finally its several species—its registral pitch (*hauteur*), its pitch character, its intensity, etc. For example, a sound object's type might be variable in mass, and it might have a "reiterating" dynamic, a "resonant" grain, and a "mechanical" allure. The class of the same sound might have a "knotty" mass, a crescendo dynamic profile, a "matte" grain, and one of three fluctuating allures. Under the first species determination, that of registral pitch, the sound might be in a middle register of mass (diapason), be unclassifiable in dynamic, have a certain "color" of grain, and be unclassifiable with regard to allure. Schaeffer's name for this process, the *solfège* of musical objects, is appropriate; it is a discipline through which one learns to listen closely and analytically, from many points of view, to musical or potentially musical sounds.

Traité belongs to a peculiarly Continental tradition that emphasizes the rationalization and philosophical justification of one's views.

Fully half the book is taken up with material of this sort which has the effect—at least for readers outside that tradition—of diluting and weakening the central issues. An argumentative, didactic, sometimes even polemical style often obscures Schaeffer's points and makes the book difficult—for foreigners, in particular. On a more substantive level, Schaeffer can be faulted for seeming to claim for himself the discovery of well-known psychological concepts—the lack of congruence between the acoustic and the perceptual (auditory) realms, for example. Important details omitted from the descriptions of the few psychoacoustic experiments that Schaeffer reports in the book make replication difficult or impossible. Since certain of the results of these experiments appear to contradict well-established psychoacoustic relationships, this is a serious problem. Schaeffer's insistence on "reduced hearing" to remove the sound object from association with the sound-producing device is inconsistent with his metaphorical appeal to the natural world for the classification process. It is hard to see the utility, for example, of first "reducing" a metallic sound and then assigning it a "mechanical" allure. The classification schemes themselves are entirely psychological: they are attempts to describe the auditory experience of a listener without specifying the acoustic processes that lead to those experiences. One would have to be naive indeed to hold that any detailed classification of the entire world of sound could be simple or straightforward. But Schaeffer's response to this complexity, his use of metaphor and dependence on what amounts to purely introspective methods, detracts from the objectivity with which he has pursued his chosen task. The reader is left with the impression that only insiders can become adept in these methods. Schaeffer's research program and *solfège* simply do not contain sufficient safeguards against prejudice and the idiosyncratic.

Nevertheless, Schaeffer's book performs a service of great value simply by exposing the complexity of these "insides" of sounds. The book effectively explodes the boundaries of pitch and rhythm within which music-theoretical study is so often enclosed. Schaeffer's background in *musique concrète* has clearly influenced him in this encyclopedic attempt to open to the consideration of composers a gigantic world of sound. And by implication the book challenges composers to find new ways of structuring this greatly expanded world in a musical context. Ingmar Bengtsson (1969), in a searching and thoughtful review of *Traité*, calls for its translation into English, suggesting that, given a wider readership, the book might well inspire new music, suggest new research into the sounds of music, and provoke insightful and instructive debate.

Robert Erickson's *Sound Structure in Music* (1975)

Erickson is much better known in America than Schaeffer, both as a writer about music and as a composer of witty and original music. His *Sound Structure in Music* covers some of the same ground as Schaeffer's book,[7] but from the opposing, "empirical" perspective of the English and American tradition. Erickson's book has been reviewed in several prominent journals (e.g., Swift 1975; Slawson 1978) so it need not be discussed in detail here.

Like Schaeffer, Erickson invents or adapts to his own purposes a series of terms to describe a variety of sonic phenomena. He discusses "fused ensemble timbres," "sound masses," "rustle noise," "spectral glide," and (*pace* Schaeffer) "grain" and "mass." Unlike Schaeffer, Erickson cites many musical examples, largely of contemporary music but including some non-Western music. *Sound Structure* offers little system or theory: the categories seem to flow from the music that Erickson is interested in, not out of an a priori rationale. We are given only a sense of the wide range of musical techniques available to present-day composers and the beginnings of a terminology for those techniques. Erickson's enthusiasm for the new and unusual is exemplified by his discussion of Ives's *Universe Symphony*:

> It is too late for Ives to hear the sounds of the music described above, but the ideas underlying that vision, multiplicity, sound-source groupings, sound perspective, sound as embedded in the natural order, sound in its physicality and its endless variety, undergird much of the significant music we are hearing today.
>
> *(p. 193)*

[7] Oddly, Erickson does not mention Schaeffer either in the text or the bibliography.

Erickson's study has been criticized (e.g., Swift, 1975) on several counts. Erickson claims to have derived certain principles from auditory theory and experiment, but he often misreads or misinterprets the scientific work he cites, and he seldom applies it in his musical analyses. The categories Erickson distinguishes are nearly all defined by example, sometimes quite casually. The discussion of "grain," for example, although highlighted by a section of its own and an entry in the Table of Contents, consists of a loose, passing reference, in the context of some observations about very fast passages of brief sounds, to the possibility "that what is important is not the grain of sounds but the grain of music, even the grain of a particular composition" (p. 75). The term is treated as though it were a standard part of the musical vocabulary. Terminological problems such as these are serious weaknesses of the book, and they limit its usefulness considerably.

Erickson's book does expose the reader to a broad range of unusual, even exotic, musical phenomena. In this it is similar to the Schaeffer work. Erickson, however, stretches the reader's conception of what can be considered a musical sound by copious references to actual music. His discussions of this music are seldom very analytic or complete, but in nearly every case we are told enough to understand the scored examples. And even in the cases of the most unusual notations and musical concepts his descriptions help us begin to imagine how the music might sound.

Robert Cogan and Pozzi Escot's *Sonic Design* (1976)

The most original chapter in *Sonic Design* is entitled "The Color of Sound." The authors' goal is to develop a method of analysis to explain "the choice and succession of the tone colors of a musical context [and] the principles that interrelate the diverse sounds of a given work" (p. 328). Such a method would be an important step toward discovering what "laws" composers may be following in structuring sound color. In contrast to the other two studies reviewed here, Cogan and Escot have little concern with classification; they are interested in how composers *organize* "tone color."

The method presented in *Sonic Design* starts with a detailed survey of the frequency spectra of the sounds made by musical instruments based on the results of, and extrapolations from, a wide variety of experiments by music-acousticians. For simultaneous sounds played by instrumental combinations, the frequency spectra of the individual instruments are simply added together to derive a combined spectrum. Connected portions of compositions and even whole pieces are analyzed by drawing successive graphs of these combined spectra, with a new graph for each change of pitch or instrumentation. Since the publication of the book, Cogan has begun to automate this process with an instrument that is somewhat similar in function to a sound spectrograph (Cogan 1980). The instrument produces a frequency-time-intensity graph through a photographic process that permits the analysis to be carried out fairly quickly. The "musical space" of a piece's "sonic design" can be assessed by examining the general shapes in these graphs.

Both the strengths and the weaknesses of Cogan and Escot's methods are demonstrated in their analysis of Debussy's *Nuages*. The large-scale tendencies in Debussy's choices of instrumentation and pitch register—resulting in a "space" covering the middle registers at the beginning of the piece, a spreading of the space toward higher frequencies, next a lowering in frequency of the still widely spread space, then a further lowering in average frequency with successive elimination of the middle registers, and finally, after an interruption that resamples the higher registers, a very restricted low-frequency space—show strikingly. The analysis is less revealing about the manner in which Debussy may have related medium and small-scale color structures to that high-level organization.

A persistent problem is posed by Cogan and Escot's failure to develop a precise and consistent terminology for speaking of the *color* of an instrument or instrumental combination, if we take that term to mean the listener's perception of the sound of the instrument or instruments. They speak of rich, dull, or bright "spectra" or they refer, more properly, to the "strength" of certain frequencies in the spectrum of an instrument. A portion of a spectrum is said to be "activated" or to move through a "spectral descent." The language is that of acoustic measurement, not—as in Schaeffer and to a lesser extent in Erickson—auditory perception. Indeed, Cogan and Escot treat their resultant spectra as more or less direct representations of the perception of the listener. Even if it can be granted for the sake of argument that the graphs of these spectra do not distort perceptual reality significantly, it is hard to develop sound-color *relations* out of these representations of entirely acoustic phenomena. Some theoretical basis for the ways listeners categorize color is required in order to begin to investigate musical regularities. Cogan and Escot do not suggest that their sound-color/musical-space graphs would serve for the study of pitch relations, although the graphs could be made to register all the fundamental frequencies in a composition. Such common pitch relations as octave equivalence or the potential for transposition or inversion in a pitch succession would be missed. If relationships of comparable subtlety among different sound colors exist or can be defined, they too are unlikely to be found in the large-scale acoustical measurements advocated by Cogan and Escot.

The originality of *Sonic Design* remains its strong point. It represents the considerable insights of two composers who have a firm knowledge of the history of Western music and who also deeply appreciate the music of other cultures. Their claim that sound color is determined by the frequency spectrum has considerable analytic power. I suspect that their own music, often frankly experimental and sometimes inspired by non-Western music, has suggested such important concepts as musical space and sonic design. Their insistence on questioning accepted ideas of musical analysis makes their study provocative; their sure musicianship ensures its lasting value.

None of these studies makes much headway toward a "theory that would determine laws" for the combination of tone colors. Schaeffer's methods of categorization appear to be based on rules, but they are not compositional rules. We are guided by the "solfège" into an unprecedented level of detail in analyzing the sounds of music, but nothing is said about the constraints those analytical categories might impose on the succession and combination of sounds in a musical context. Erickson urges us to listen in more detail and illustrates his observations with sounds in real compositions. But very little is said about any general principles that may have constrained the juxtapositions of those sounds. His book is not about rules at all. Cogan and Escot's methods are mostly descriptive. Large-scale organization of tone color is indicated by their graphs, but those graphs are incapable, without further interpretation, of revealing systematic methods for connecting individual sound colors into the small and middle-level structures from which those larger structures might be built. Even the observations made about the large-scale structures refer only to rather vague esthetic criteria, not to methods of composition with sound color. All these studies, thus, have terminological and methodological problems, and none gives substantial guidance for the combination of tone colors into a coherent musical structure.

A theory of sound color should provide such guidance. But what form would such a theory take? What kinds of questions should it be capable of answering? To what extent should it be restrictive, even prescriptive, and to what extent only descriptive? By addressing these fundamental issues directly, perhaps some of the shortcomings of the previous studies of timbre can be avoided.

FORMAL REQUIREMENTS OF SOUND COLOR THEORY

Musicians, more than practitioners of any of the other arts, are prone to theorize. Historically, the results of these systematic examinations of music and musical craft have taken many forms. Some scholar/musicians have identified a hitherto unrecognized musical practice, have formalized it to some degree, and then have attempted to show that their formalization explains some aspect of the respected music of their own or some previous time. Glarean's *Dodecachordon* is an example of this kind of theory. Others—Morley and Heinichen come to mind—have written manuals of instruction in composition that contain generalizations, either explicit or implicit, about musical structure. Still other writers—Rameau, for example—have appealed to mathematics or some other extramusical system to justify their music-theoretical claims.[8] In all these important studies of musical processes certain

---

[8] Two of the theoretical works referred to here are available in English translation (Glarean 1965 and Rameau 1971). The Heinichen and Morley books have been reprinted in facsimile (Heinichen 1969 and Morley 1953).

meta-theoretical issues are, justifiably, taken for granted. However, studies in the new area of sound color, which has little well-developed theory, must be more explicit about these fundamental issues: the definition of theoretical elements, the psychoacoustic correlates of the elements, the distinction between invariance and difference in the questions the theory attempts to answer, and the explicit definition of musical operations.

Elementary Units

Most music theory in the Western European tradition has concerned pitch. At least one reason for this focus is the general agreement about what constitutes a single pitch and when a musical voice or line moves on to another pitch. We must recognize an occasional concern with tuning, but for the most part individual pitches are taken as givens. Thus, there is no necessity to define to what elements pitch theories are to apply. There is no agreement, however, about what constitutes an element of sound color. Is sound color to be associated with a specific musical instrument, say, a particular Stradivarius violin? This meaning is suggested when we say that an instrument has "a good tone." Or should the sound of all instruments of the same type be taken as the basic element of color? When the violin, any violin, is said to have a different "instrumental color" from the oboe, we are using the term that way. Should we include sounds not usually thought to be musical, as in speaking of the "timbre" of a glass breaking? Alternatively, should we speak of timbre or sound color in more abstract terms, after "reducing," in Schaeffer's terms, what we hear? Clearly, the first requirement of a theory of sound color is to choose what musical elements the theory is to deal with.

Psychoacoustic Definition of Elements

An unstated assumption in pitch theories is the relation between the elements of the theory—the pitches—and the physical actions—the rapid changes in sound pressure in the air—that give rise to them. This is not to say that the psychoacoustic relations underlying the sensation of pitch are completely understood (see, for example, Shepard 1982). But musical, as opposed to psychoacoustic, theories of pitch are not concerned with these matters, because in the broad range of sound usually considered to be musical space, the relation between frequency and pitch is well known and uncontroversial.

But a relationship that can be taken for granted when discussing pitch must be specified in the case of sound color. If we were to choose to theorize about the color of musical instruments, we might avoid having to define the elements of our theory in acoustic terms.[9] But if we want to speak of the color of a sound as a psychological dimension—divorcing the sound from its manner of production—we can eliminate the possibility of introducing bias (or the need for special knowledge that weakens Schaeffer's approach) only if we are able to match our observations about what is perceived to the acoustic events that give rise to those observations. In other words, the theory should have a psychoacoustic component.

Invariance versus Difference

Another issue has to do with the kinds of questions the theory is to attempt to answer. Almost all music theories about pitch assert some relationship that is *invariant*. Thus, Rameau's foremost claim was the assertion that chords in inversion—those figured "6" and "6–4"—belong in the same category as the parent chord in root position. The chord's fundamental bass is held invariant when the chord is presented in inversion. Recent theories of atonal music are quite explicit about defining invariances; permutation of the order of a set of pitch-classes, for example, leaves invariant the interval-class content of the set.

It may seem that this preference for hypothesizing about invariances springs from an esthetic stance that holds invariance in music to be a more important organizing force than "variety" or "going on to something new."[10] To some degree this stance must be recognized

---

[9] Other, far more difficult, problems would arise in this case. For example, musical instruments are notoriously adept at changing their color with changes in pitch (see discussion of musical instruments in Chapter Two).

[10] Stravinsky (1956) has expressed this view eloquently.

as a bias, but it is an attitude that is widely held. All of science requires statements in this form. Seldom is the demonstration that two phenomena are different greeted with any enthusiasm, but when many apparently different phenomena are shown to obey the same law, publication in prestigious journals and professional recognition are earned.

According to this argument, the theory of sound color should provide answers to questions about how one can hold sound color invariant. Clearly it would be desirable to specify how to preserve color under changes in the loudness or the duration of a sound. We would like to know how to change pitch without changing sound color. Moreover, if sound color itself is a complex phenomenon made up of several different aspects or dimensions, we must show how one aspect of color can be held invariant as other aspects are varied.

All three of the studies reviewed above make implicit claims of invariance. Schaeffer's categories bring together different sounds that share certain traits. Erickson's lists of phenomena suggest that different pieces share certain compositional procedures. Two pieces that have similar graphs according to the scheme of Cogan and Escot can be said to have the same or similar color structures. But the emphasis in all three studies, it is fair to say, is on differences. In each case we are asked to hear—indeed, to learn to open our ears to, and marvel at—the great variety of musical sound. Very little is said explicitly about how different sounds are related to each other. The emphasis in this study will be quite in contrast; here all the theoretical statements will be in the form of assertions about the invariance of sound color in the face of changes of various sorts.

Operations on Sound Color

The permissible operations on musical elements are a central concern of nearly all music theories. The operation of transposition is a prominent and well-developed feature of the music of the "common practice" period. In the simplest form of transposition, a melody is rewritten starting on a different but tonally related scale step. Inversion of a melody is another operation employed by European composers from the Renaissance to the present. Versions of these operations were used by Schoenberg and his associates and have been investigated formally by composers and theorists (Babbitt 1960; Forte 1964). All these operations are means of maintaining invariance. If, for example, a sequence of pitches is inverted, the transformed sequence will in most cases differ from the original, but important features of the sequence will be retained. No Western music—possibly no music at all—would be possible without operations of some kind. They are the basis for all variation, that is to say, all ways of holding constant some aspects of the music while changing others.

Studies of sound color conspicuously lack definitions of operations. Yet, if Schoenberg's vision of a logical method for organizing sound color is to be pursued, we must identify operations that transform sound colors while holding some aspect of those colors invariant. It would be interesting to a composer if the color operations could be related somehow to pitch operations. But the pitch operations should not be slavishly imitated in defining operations in the color realm. They should be derivable in some natural way from the properties of sound color itself. While, ideally, a psychoacoustic justification for the operations should be found, this is not a critical issue. Indeed, musically productive operations on pitch often have rather complex psychoacoustic underpinnings. It would be hard to argue, for example, that an inverted melody is identified by the kind of low-level auditory process usually investigated by psychoacoustic methods. Sound-color operations that require high-level cognitive processes and learning should not be rejected out-of-hand if there are other good reasons for including them in the theory.

TIMBRE VERSUS SOUND COLOR: LIMITING THE SCOPE

We have now set demanding and far-reaching goals for the theory of sound color. The next step is to delimit the domain over which the theory is to be applied. The theory (stated in Chapter Two) will provide a definition of that domain; here we can develop a preliminary notion of what the bounds of application are and make a terminological distinction that will help to keep those bounds in mind. Let us consider some of the many terms that have been used to refer to that unexplored country that so concerned Schoenberg.

Timbre, according to the American Standards Association (1960), is the term covering all ways that two sounds of the same pitch, loudness, and apparent duration may differ. We do not need Schaeffer's treatise, but only a moment's reflection, to convince us that timbre, understood in this way, encompasses an enormous variety of phenomena. Even the sounds of musical instruments, with their various attack transients, the variations within the "held" portions of notes, and their many decay characteristics, present a bewildering array of disparate acoustic events. But the domain of timbre would have to include most of the categories in Schaeffer's treatise that pertain to non-instrumental sounds as well. The multitude of acoustic variables involved before perception of timbre could be considered removes any hope of writing a theory that meets the requirements outlined above. Either the detailed specification of invariances and operations that preserve them must be given up, or a smaller realm must be delineated over which the theory is intended to apply. Schaeffer chose to be encyclopedic, thereby limiting himself almost entirely to categorization, with little discussion of invariance and none about operations. In this study the second alternative is taken: the theoretical claims are limited to a small, well-defined subset of timbre.

What should the subject of the theory be called? One could simply criticize the American Standards Association for offering only a catch-all, and then redefine *timbre* to correspond to the intended domain of the theory. But since the term is already used widely in many different senses by scientists, music theorists, and composers, it would be hopeless to try to reach agreement on a single meaning. In any case, it is often convenient to be able to speak generically about, say, the range of phenomena Schaeffer discusses.

There are other options. Any term that includes reference to a specific mode of sound production should be avoided if we are to keep the requirement that the theory address sound in the abstract. This rules out *instrumental color, instrumental timbre, vowel color, vocality*, etc. Another possibility is *tone quality*, but terms that include the word *quality* imply a value as in "that trumpet's bad tone quality," and the term often is used to refer to the sound production of a particular player or singer, and terms that include *tone* are prone to be confused with references to pitch.

Sound color, or simply *color* when the context makes clear the reference to sound, avoids the drawbacks of the other terms. It is sometimes used by musicians, but no official meaning has been established as for *timbre*, and there is, to this point, no commonly accepted, precise usage. One disadvantage is the term's association with the phenomenon of synesthesia, that is, the tendency of certain sounds—sometimes even certain pitches—to suggest to some people a visual sensation of color (see, for example, Marks 1978). This, of course, is not intended in the present study, although the analogy with visual color helps to suggest the auditory meaning.

Sound color is a property or attribute of auditory sensation; it is not an acoustic property. Similarly, visual color is a perceptual attribute, not a property of light. Sound color, like visual color, is abstract; no specific source of energy is implied by either term. Again like visual color, sound color has no temporal aspect. A light may be described as fluttering rapidly between two colors or slowly changing in color, but the flutter or changing shade is not itself regarded as a color. When we say that a sound color has no temporal aspect, this rules out of consideration all changes in sounds.[11] That is, a sound may be heard to be changing from one color to another, but the change itself is not a sound color. It follows that sound color pertains to the steady-state portions of sounds but not, in general, to their beginnings or endings, where we can sometimes hear rapid changes in the character of the sound. Sound color and visual color are multidimensional, both may be mixed, and both are prominent, quite general properties of sensation.

THE PRINCIPAL QUESTIONS

In accordance with the theoretical goals outlined above, this study addresses three basic questions:

1. How can the color of a sound be held constant when its pitch, loudness, and other aspects of its timbre are changed?

[11] It may be convenient to discuss such changing sounds as having a certain timbre—some degree of Schaeffer's "grain," for example. J. K. Randall (1967) has criticized the application of the term *timbre* to such changing sounds, however.

2. How can one aspect or dimension of sound color be held constant as other dimensions of sound color are varied?

3. What operations on sound color can be identified that will hold invariant certain relations among sound colors while varying other relations?

If these questions can be answered successfully, a large step will have been taken toward codification of the "laws of tone color" that Schoenberg found wanting, and a foundation will have been laid for a new way of structuring music.

CHAPTER TWO

A Theory of Sound Color

IN THIS AND THE REMAINING CHAPTERS of this study attempts will be made, from several points of view, to answer the central questions posed at the end of Chapter One. The approach in the present chapter is to propose theoretical answers, arguing from rational premises rather than empirical data.

THE SOURCE/FILTER MODEL: THREE EXAMPLES

The primary basis for the theory of sound color I am proposing is a conception of how certain sounds are made: *the source/filter model* of sound production. A broad class of sounds can be analyzed productively with this model. To introduce the model, to indicate to what kinds of sounds it can and cannot be applied, and to answer in a preliminary way the first of the central questions of this study, let us consider in turn three contrasting examples of familiar sounds.

The Wall Tapper

Those of us who have indulged in the renovation of an old house are likely to have spent some time tapping on walls and ceilings. This peculiar occupation is an attempt to locate the supporting joists behind the plaster or wallboard. We have to find these elusive structural members before cutting holes to mount lights, electric sockets and switches, etc. As we tap with our fingernail along the wall, we listen for the telltale rise in "pitch" that indicates a stiffer wall and, thus, the probable presence of the underlying two-by-four. The acoustical foundations of this unfortunately not-always-successful procedure can best be understood in terms of the source/filter model.

According to this model, a sound is produced when an object is struck, or *excited*, by some kind of mechanical energy and the object in turn changes the excitation in some manner. The mechanical excitation is called the *source*; the object itself, the *filter*. In the case of the house renovator, the tapping fingernail is the source and the wall or ceiling is the filter. Notice that the source and filter have quite distinct and independent characteristics. The source intensity and its timing—one may prefer fast or slow tapping or one may stop altogether when someone turns on the phonograph loudly—are all under the control of the renovator. The "pitch" of the sound that results from tapping the filter, however, is a property of the particular section of wall. The independence of the source and filter and, at the same time, the modification of the source by the filter are the essential features of the source/filter model. These features characterize our first example, for the wall can be said to modify the energy that strikes it from the independently tapping finger. Both parts, source and filter, are necessary: without a source the wall is mute, and without the wall's modification of the source energy we can learn nothing about its local stiffness.

This model and example set the stage for a partial answer to the question of how to keep the color of a sound constant while changing other characteristics. Let us call this version of the answer *Rule 1'*:

> *Rule 1'*: Sound color is associated with the filter, not the source. To keep sound color constant, keep the filter constant.

Applying Rule 1' to the first example, we can interpret the "pitch" of the sound coming from the filter (the wall) as a sound color. Tapping harder or more rapidly or with different objects will change certain characteristics of the sound, but unless we tap at a different place on the wall or on some other object entirely, the sound color will not change. Because the house renovator wants information about the filter, he or she tries to keep the tapping as constant as possible while moving across the wall or ceiling listening for the change in sound color that signals a difference in the stiffness of the wall. Rule 1' can be thought of as a definition of sound color as "belonging" to an object which is silent until it encounters energy from some exterior source. This rather odd state of affairs means that we hear sound color only indirectly, when "something else" happens to the filter.

Vowels in the Human Vocal Tract

Our second example, from which much of our theory of sound color is derived, is the human voice pronouncing vowels.[1] The main source in vowels is the action of the vocal chords, which move much faster than the tapping of the fingernail. The main filter is the cavity formed by the throat and mouth (see Figure 2). The source/filter model, which was invented to deal with speech, is a particularly good one in this situation. We all know that we can convert a declarative sentence to a question by changing the pitch pattern to a rising inflection:

You went home.   You went home?

The movements of our mouths and throats are the same in these two sentences; transformation into a question is brought about by changes in the action of the vocal chords. Since the filter patterns are the same, then, according to Rule 1', the sound colors must be the same. To complete the picture let us consider two different sentences that share an inflectional pattern:

You went home?   He flew where?

Here the vocal-chord patterns in the two sentences are very similar but the filter movements, and hence the sound colors, are quite different.

[1] Consonants are an important part of connected speech, of course. In general they have other sources—e.g., noise arising from turbulence at various places in the vocal tract—and other filters—e.g., the nasal passages. For our present purposes we can set aside these more complicated actions of the vocal mechanisms.

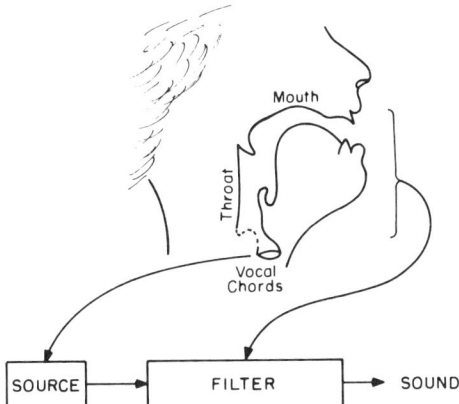

FIGURE 2: *The source/filter model of the vocal mechanism.*

The source/filter model appears to distinguish rather neatly between the phonetic aspect of speech—the filter—and the inflectional or stress aspect—the source. Actually the distinction is not quite so clear-cut—it is contradicted in tonal languages, for example—but as a first approximation the model is useful even at this rather high linguistic level. Application of the model has been most fruitful in the detailed analysis of individual speech sounds. Let us examine how it represents the utterance of vowels.

Suppose we position our vocal tract so as to produce the vowel [a] (as in "rod"), and tense our vocal chords so as to produce the pitch A below middle C (fundamental frequency 220 Hz). We can represent this steady-state situation with the frequency spectrum in Figure 3.[2] The source spectrum (somewhat idealized) is harmonic, with partials at 220 Hz, 440 Hz, 660 Hz, etc., that decrease regularly in amplitude as the partial numbers increase. The kind of modification the vocal-tract filter imposes on any source can be represented as a template that attenuates some frequencies and reinforces others. This template is called the *spectrum envelope*. In the spectrum envelope of the vowel [a], the frequencies in the neighborhoods of 800 Hz and 1200 Hz are emphasized, those well below 800 Hz are altered only slightly, and those above 1200 Hz are sharply attenuated. This pattern of hills and

[2]For an introduction to the concepts of *frequency spectrum, fundamental frequency*, etc., that leads naturally into the present discussion see Slawson (1975).

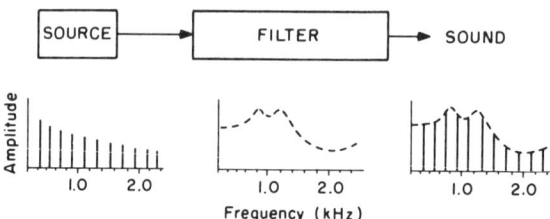

FIGURE 3: *The source and filter in utterance of the steady-state vowel [a] at a source frequency of 220 Hz.*

valleys "belongs to" the vowel [a] just as, in the case of the house renovator, a characteristic spectrum envelope that "colors" the tapping vibration "belongs to" each position on the soon-to-be-violated wall.

The spectrum of the sounded vowel contains information about both the source and the filter. The amplitudes of the source components are modified, but their frequencies are not. The pattern of the amplitude modifications provides a representation of the spectrum envelope and, by implication, the characteristics of the vocal tract, for the sound. The representation is not perfect—only a few points on the spectrum envelope are reflected in the pattern of the source's line spectrum—but an indication of the filter's characteristics is given.

We can identify several steps in the process of uttering a vowel. We begin by positioning our vocal tract appropriately and causing our vocal chords to vibrate. A characteristic spectrum envelope is imposed on the source spectrum, and finally a sound is produced in whose spectrum the spectrum envelope is represented more or less sparsely. To complete the communication process, the auditory system of the listener must then perform a series of analytic operations.

By associating sound color with the filter characteristic, Rule 1' claims, in effect, that the listener's auditory system attempts to learn something about the filter by analyzing the color of the sound. We are in a position now to propose a more useful version of that rule in terms of the spectrum envelope:

Rule 1″: To keep the color of a sound constant, keep its spectrum envelope constant.

Suppose, for example, we want to lower the pitch of the vowel by the interval of a fourth, to E (fundamental frequency 165 Hz), while keeping its sound color constant. Rule 1″ says that we must keep the spectrum envelope constant as we change to the lower, and thus more closely spaced, source spectrum (see Figure 4). We notice that although the spectrum envelope reinforces the same frequency regions in the lower-pitched vowel as it did in the vowel pitched a fifth higher, different components of the source fall within those regions. Thus in the first vowel the fourth partial is reinforced, whereas in the second vowel, the fifth partial is closest to the highest "hill" in the spectrum envelope.

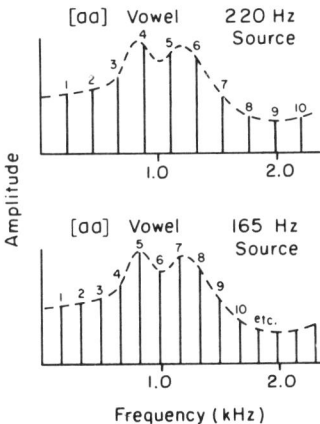

FIGURE 4: *Frequency spectrum produced by holding sound color constant while lowering the pitch.*

Rule 1″ is really only a more precise version of Rule 1′. Where Rule 1′ tells us to keep the filter itself constant, Rule 1″ specifies that we should keep the *effect*—that is, the spectrum envelope—of the filter constant.

Vowels and vowel-like sounds will be discussed in considerable detail below, but let us stop at this point and ask what reasonable alternatives there might be to Rules 1′ and 1″. Our third example, musical instruments, will suggest one.

Musical Instruments and "Strong Coupling"

The instruments of the orchestra produce sounds in a variety of ways. With the exception of some percussion instruments, however, they all share a mode of action which contrasts with that of the wall being tapped by a fingernail or the vocal tract being excited by the vocal chords and which fails to meet one of the requirements of the source/filter model. All the string and wind instruments have a source (the bow or a buzzing reed), and all have a filter (the string or the horn). Why doesn't the model work? The answer lies in the way in those instruments the source interacts with—technically, is *coupled* to—the filter.

Consider the clarinet. We can buzz at any pitch on the mouthpiece alone. However, when the mouthpiece is inserted into the horn, only the pitch that has been fingered can be played.[3] If we attempt to buzz at A below middle C (220 Hz), but we have covered the holes in the instrument in such a way as to produce G a whole step below, the G will sound rather than the A. The source may start the sound at its own frequency but immediately the filter, at *its* favored frequency, begins to act on the source, forcing a change. This feedback from the filter to the source changes the picture of the source/filter model (see Figure 5). The source is *strongly coupled* to the filter in the clarinet. That is, the clarinet source is not independent; it is driven by the filter. In vowel production the filter changes the amplitudes of the source components but does little else to the source; source and filter are *weakly coupled*. In the clarinet and most other musical instruments, the filter affects the source frequency as well. In fact, the usual method of changing the pitch of a musical instrument is to alter, not its source characteristics, but the effective length of the horn or the string—in other words, the filter.[4]

Figure 6 and Sound Example 1 illustrate this point by comparing the effects of two hypothetical filter systems, one weakly coupled to its source and the other strongly coupled. Let us suppose that both systems happen to have the same spectrum envelope when the sources are

---

[3]Without heroic efforts, that is. The exceptions—glissandi, multiphonics, and other techniques, some of which are cited in Erickson (1975)—are valuable extensions of the range of sounds available to the clarinettist. I am speaking here only of the standard manner of playing this and the other instruments I shall cite.

[4]The analysis of the stringed instruments is similar to that of the winds if the action of the bow hairs is treated as the source and the string is treated as the filter.

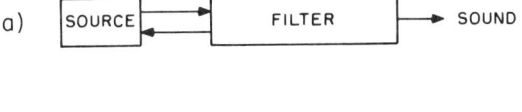

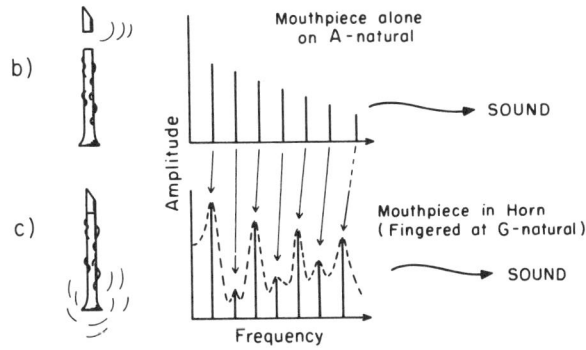

FIGURE 5: *Source and filter in a clarinet. (a) A feedback path is added to the source/filter model. (b) The mouthpiece, detached from the horn, is buzzed at A♮ with the horn fingered at G♮. (c) The components from the mouthpiece are forced by the filter to sound at G♮.*

at about 65 Hz—the pitch of the open C string on the cello. Now we can ask what the spectrum produced in each system would be at a pitch an octave higher. In the weakly coupled case, the filter—thus the spectrum envelope—stays the same, emphasizing and attenuating a set of the more sparsely distributed partials of the source in a pattern quite different from the pattern at the lower pitch. In strongly coupled systems, on the other hand, the filter causes changes in pitch, so the filter itself must change. The spectrum envelope will stretch out to higher frequencies, with the result that the partial amplitudes remain the same as they were at the lower fundamental frequency. [RECORDED EXAMPLE 1A] [RECORDED EXAMPLE 1B]

What can we say about sound color in strongly coupled systems? Rule 1' and 1" do not apply becaue there is no way, in such systems, of controlling source and filter independently. Some alternative to those rules would have to be proposed. For example, we might formulate the following:

> *Rule X:* To keep sound color constant, keep the relative intensities of the partials constant.

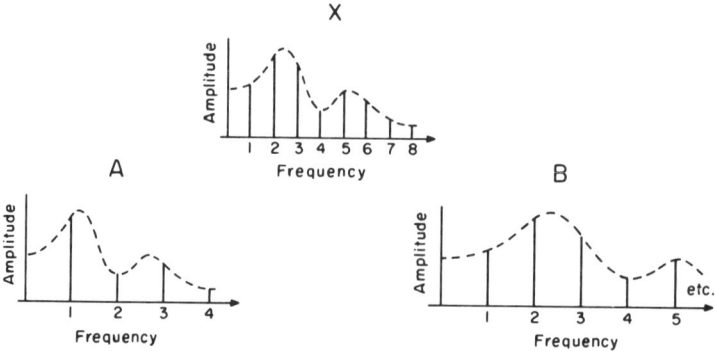

FIGURE 6: *Changing pitch under weak and strong coupling. If the spectrum X is produced by a source/filter system that is weakly coupled, then raising the pitch by an octave will result in A. If X is produced by a system with strong coupling, the filter itself changes with the change in pitch. If we assume that the only change in this filter is a proportional "stretching," then raising the pitch by an octave results in alternative B—a new spectrum envelope.*

Figure 6 emphasizes the difference between the definition of color implied by this rule and that of the rules that grew out of consideration of systems in which the source and filter act independently of each other.

How are we to choose between these contradictory views of sound color? Perhaps the best way is by carrying out psychoacoustic experiments to see which alternative is favored by human listeners. The results of experiments designed to do just that will be reviewed in Chapter Four; here it is enough to say that the preponderance of the evidence favors Rules 1' and 1". It is hard, however, simply to reject Rule X, for it grows out of an analysis of the way musical instruments function, and the term *color* has been used by many musicians in reference to the sounds of musical instruments. This dilemma may be resolvable by closer analysis of characteristics of musical instruments that do not change with pitch. In other words, we can look for subsystems of a musical instrument that are weakly coupled to the rest of the instrument and that may be largely responsible for the sound color of the instrument. The beginnings of research along this line will be reviewed in Chapter Five.

There remain acoustic systems that do not fit the source/filter model very well: some perceptual regularity, such as Rule X, may hold for sounds produced in those systems. But should whatever Rule X applies to be called *sound color*? Perhaps a better course would be to find some other term, such as *instrumental timbre* or *instrumental quality*, for this possible psychological attribute that so differs from sound color as specified in Rules 1′ and 1″. Let us save *sound color* for cases in which the source and filter are weakly coupled and it makes sense to associate a psychological quantity with the distinguishable part of an acoustic system. Musical instruments, to the extent that they behave as strongly coupled systems, are devices in which sound color cannot be held invariant as the pitch changes. They illustrate one kind of limitation of the domain to which this study is intended to apply.

A Cautionary Note

The foregoing simplified analysis ignores certain features of each of the acoustic systems. The sources in the first two examples are actually complex systems that include filter-like phenomena, and musical instruments are more complicated than the discussion of example three suggests. Specifically, the coupled source/filter system of many musical instruments can be analyzed as a particularly complex source that excites certain fixed resonances in, e.g., the bells of horns or the bodies of string instruments.[5] We shall return to some of these issues below.

RESONANCES AND THE SPECTRUM ENVELOPE

Although somewhat more revealing than Rule 1′, Rule 1″ leaves questions unanswered. What causes the "hills" and "valleys" in the spectrum envelope? What is the relationship between the spectrum envelope and the physical properties of the filter? Are there parts of the spectrum envelope that can be used to specify the whole envelope? These questions can be answered by a deeper analysis of how filters function, yielding the final version of Rule 1, upon which the rest of the theory depends.

---

[5] Additional fixed-plate and cavity resonances in stringed instruments are discussed by Jansson (1966). Fixed resonances in wind instruments are discussed by Fransson (1966) and by Strong and Clark (1967).

Simple Resonance: The First "Hill"

Weights on a spring, narrow-necked bottles, and some kinds of bandpass filters in the electronic music studio—all these devices require external energy to excite them and they in turn modify that external energy in some manner. These filters are all *simple resonators*.[6] The form of their spectrum envelopes is illustrated in Figure 7. Even though the spectrum envelope of simple resonance—sometimes called the *resonance curve*—may not seem particularly simple, it can be specified entirely by two quantities, the *center frequency* of the resonance and its *bandwidth*. The center frequency is the frequency of the highest point on the spectrum envelope.[7] This locates the "hill" on the scale of frequencies. The bandwidth specifies its steepness, how high it is above the surrounding "plain." The bandwidth is measured, by convention, at the two points on the slopes of the curve that are three decibels lower than the peak and is defined as the difference in frequency between those two points. If the resonator is finely tuned, the peak of the curve will be very narrow and high in relation to the rest of the curve, so the bandwidth will be small. So-called dull resonators have a broader, less pronounced peak and large bandwidths. The center frequency of a resonance is often called, naturally enough, the *resonance frequency*. Other names for it are *formant frequency, pole frequency, cut-off frequency*. The reciprocal of the bandwidth is called the Q or *quality factor*: narrow, high peaks have a high Q; broad, low peaks, a low Q.

While it may seem reasonable that the region of the resonance

---

[6] These devices must be idealized to satisfy the conditions for simple resonance. In fact no object in the real world is a simple resonator, but the action of some of them, such as the three examples cited here, can be closely approximated by assuming that they are simple.

[7] The center frequency is most unambiguously defined in terms of phase relations between the source and the response of the filter. A clear, rigorous exposition is given by Fant (1960). This exposition of filter theory assumes system linearity. In other words, the amplitude of the source must be small enough that the only effect of the resonance system is to multiply the amplitude of all components of the source by the value of the spectrum envelope at those components' frequencies. There must be no scattering of energy due to the source overdriving the filter.

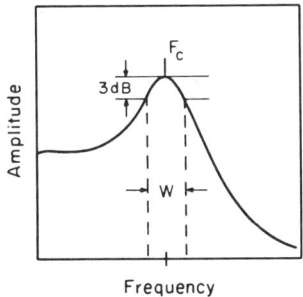

FIGURE 7: *Spectrum envelope of a simple resonator. F_c is the center or resonance frequency; W is the bandwidth, which is equal to the difference in frequency between the low and high frequency "skirts" of the curve at points 3 dB below the peak amplitude.*

curve in the vicinity of the peak can be specified by the center frequency and the bandwidth alone, our intuition suggests that the rest of the curve will require additional specification. In fact it does not. All simple resonators are similar in their effects on frequencies remote from the center frequency. Signals at frequencies considerably lower than the peak are transmitted with little modification. Signals at frequencies above the peak are more attenuated the higher the frequency. At frequencies well above the resonance frequency, the rate of attenuation is twelve decibels per octave. These statements about the spectrum envelopes of resonances hold for all resonance frequencies or bandwidths.

It is convenient to be able to refer to the amplitude of the resonance curve at the center frequency. The *resonance amplitude* or *peak amplitude* is measured in decibels relative to the amplitude at low frequencies. It is not an independent quantity, however, but a function of both the center frequency and the bandwidth of the resonance. As a general rule, the peak amplitude is most strongly related to the reciprocal of the bandwidth. A narrow bandwidth results in a high peak amplitude; a wide bandwidth, in a lower peak amplitude.

A good way to gain insight into the properties of simple resonators is to make one and play with it. Some physics instructors demonstrate the phenomenon with something like a model car riding along a track attached to a "source" with a spring. For those who do not have access to such a demonstration, an experiment with a small hammer on a fairly stiff rubber band hanging from one's forefinger can be illuminating. If you move your finger up and down very slowly—the low-frequency case—the hammer follows at about the same amplitude as your finger. As you increase the frequency of the source (your finger), the hammer begins to move up and down at greater and greater amplitudes. (Be careful: the amplitude of the source—your finger—usually must be decreased markedly at this point to avoid catastrophe!) Finally, an ideal frequency is reached at which the hammer and rubber-band system is vibrating at its maximum amplitude—and, incidentally, always 180 degrees out of phase with your finger. At this ideal frequency, very little energy need be put into the system to get a spectacular result, and the resonator seems to cooperate with your finger. This is the resonance frequency of the system. The demonstration is completed by moving your finger even faster than the resonance frequency. As you increase the source frequency, the hammer stops vibrating almost entirely, the rubber band seeming to absorb all the energy going into the system. Acoustic resonators, whether they are bottles or tubes or cavities, all act in principle just like the hammer and rubber-band system.

This analysis applies to the simple resonator whether the source is strongly or weakly coupled to the filter. The feel of the hammer and rubber-band system at resonance suggests how filters capture sources, driving them at the filter's favored frequencies. We have to force our finger to move at frequencies away from resonance in order to investigate the filter's response to those frequencies. When the source is not independent enough to prevent the resonance from driving it at the resonance frequency the system is strongly coupled; when we move our finger faster or slower than the resonance frequency we are asserting the independence of source and filter, and the system is weakly coupled.

One common example of the simple resonator is the electronic-music-studio band-pass filter. This filter is used often in electronic music to emphasize certain frequencies from the spectra of a variety of source signals. Many such filters permit manipulation of the Q, or resonance, of the filter. All of them give the composer control of the center frequency—which is actually the resonance frequency, the frequency above which the source signal is reduced in amplitude.[8] In these filters the source can be any signal; it is clearly independent of the filter. When composers of electronic music speak of the effect of the filter as "coloring" the source, they are using the term in a way that is consistent with the definitions of sound color in Rules 1' and 1". In Chapter Six specific examples of this use of filters will be discussed.

Multiple Resonators: Uniform Tubes

Much of our considerable knowledge of the acoustics of tubes grew out of a natural, and long-standing, curiosity about our vocal mechanisms. A long line of research by scientists and engineers[9] culminated in 1960 with Gunnar Fant's *Acoustic Theory of Speech Production*. In this comprehensive, mathematically rigorous, but remarkably readable study, Fant summarizes earlier work in vocal tract acoustics and considerably advances our knowledge, particularly of the acoustics of vowels. Fant's model of the vocal tract as a compound tube serves as the basis for the following discussion.

Resonances in a Uniform Tube

Let us begin by examining a tube of uniform cross-section that is closed at one end. If we take that tube to be just 17 centimeters long it will serve as an excellent model of the adult male vocal tract during the utterance of the neutral vowel [ə], as in unstressed "the." One way of analyzing this kind of acoustic system is to ask what happens to parti-

---

[8] In most synthesizers used in electronic music, band-pass filters strongly attenuate low frequencies as well as high frequencies, whereas the simple resonator "passes" energy below the resonance frequency, although at an amplitude usually considerably lower than the amplitude at resonance. In such synthesizers the low-pass filter, with the Q or resonance set high, may approximate the behavior of the simple resonator more closely than the bandpass filter does.

[9] A few important studies of speech in the present century are Stumpf (1926), Chiba and Kajiyama (1941), and Stevens and House (1961).

cles of air at the closed and at the open end of the tube.[10] At the closed end, particle movement is constrained by the end wall of the tube and the energy of movement is converted into compressions or rarefactions (regions of minimum pressure). At the open end, just the opposite is true; the particles can move freely in and out of the mouth of the tube and pressure differences are minimized. What kinds of sounds can best be sustained in such a tube?

One answer is: sounds at wavelengths for which the distance between some particle movement minimum and some particle movement maximum equals the length of the tube—no matter how many minima or maxima may fall in between. In the case we are considering, a maximum of particle displacement 17 centimeters away from the nearest minimum of particle displacement will fit the tube very well. A sound of 68 centimeters wavelength—which, under ordinary conditions of temperature and pressure, has a frequency of about 500 Hz—satisfies this condition. But wavelengths of 22.7 centimeters, or 1500 Hz, and of 13.6 cm (2500 Hz) and 9.6 cm (3500 Hz) also fit the tube (see Figure 8). This "quarter-wavelength" tube favors not just a single frequency region as the simple resonator does, but all frequencies whose wavelengths fit the tube's boundary conditions. The tube is thus a device with *multiple resonances*.

How to "Add" Resonance Curves

The uniform tube is like the simple resonator in that both respond strongly to source frequencies near resonance but, unlike the simple resonator, tubes have an infinite number of resonances.

To draw the spectrum envelope of such a system, we treat each resonance in a multiple resonator as if it were an independent simple resonance. Then we multiply the spectrum-envelope values of each resonance together at every frequency. As a practical matter we can select a few frequencies, perform the multiplication, and draw smooth curves between these calculated points. Since we are multiplying the

---

[10] For a more detailed but non-mathematical treatment of resonance in tubes see Slawson (1975).

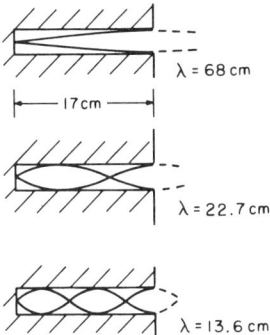

FIGURE 8: *Particle displacement in a uniform tube closed at one end. The first three waveforms that fit the quarter-wavelength tube are illustrated.*

curves, their absolute amplitudes (at low frequencies) are not important; the final shape will be the same no matter what the amplitudes are. It is convenient, therefore, to assume that all resonance amplitudes at zero frequency—their *DC amplitudes*—are equal to one. Figure 9 illustrates the process for three resonances.

The overall spectrum envelope that results from these operations has certain predictable features. There are "hills" in the spectrum envelope at the resonance frequencies of the individual resonances. Moreover, the general effect of the spectrum envelope on a source is like that of the single resonance: source components at low frequencies are changed very little, whereas at frequencies above the resonances, source components are increasingly attenuated.

The spectrum envelope of the compound resonator exhibits other features that are somewhat more subtle. The resonance with the lowest frequency has a peak amplitude that is considerably higher than that of the second resonance, the second resonance is considerably higher than the third, and so on. Successive attenuation of the higher resonances is a consequence of the steep slope on the high-frequency side of the individual resonance curves. This same high-frequency "falloff" of the individual resonances causes the increase in the steepness of the curve above the third and last resonance in Figure 9. If, however, we had included *all* the resonances of the uniform tube, the fall-off

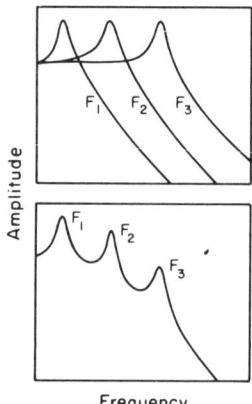

FIGURE 9: *Derivation of the spectrum envelope of a three-resonance filter (after Fant 1960).*

with frequency of the spectrum envelope as a whole would have been 12 decibels per octave, just as in the simple resonance.[11]

Formants

The term *formant* is widely used to designate the effects on the spectrum envelope of resonances in speech sounds. First coined around 1900, the term was originally used to refer to bands of relatively high intensity in the frequency spectra of vowels. Only after World War II was this rather vague usage given a specific acoustic foundation (Stevens and House 1961). Now, by universal agreement, the term is synonymous with resonances in the vocal tract and the formants are numbered from the lowest in frequency on up, with their frequencies abbreviated as F_1, F_2, F_3, etc., and their bandwidths as B_1, B_2, B_3, etc. In the neutral vowel, for example, the first formant, F_1, has a frequency of 500 Hz, F_2 is 1500 Hz, and F_3 is 2500 Hz. As a general

---

[11] It follows that the simulation of a uniform tube resonator using a limited number of simple, band-pass filters requires some kind of "correction" at high frequencies to compensate for the effects of the missing higher resonances. See Rosen (1960) for details.

rule, formant bandwidths in vowels are about ten percent of the formant frequencies (Fant 1960).

Spectrum Envelopes of Non-Uniform Tubes

The spectrum envelope of the vowel [a] ("r*o*d"), the second example cited at the beginning of this chapter, differs considerably from that of the neutral vowel. Not surprisingly, the tube shape of [a] differs as well. We can get some insight into the difference by pronouncing both vowels and noticing the differences in tongue and jaw position. To go from [ɜ] (unstressed "th*e*") to [a], we have to lower the jaw, opening the lips to their widest extent in speech. No longer are we dealing with a tube whose cross-section is even approximately uniform. Rather, the front part of the tube has a much larger diameter than the back part. This change in tube shape has a marked effect on the spectrum envelope. Taking the neutral position (F1 at 500 Hz and F2 at 1500 Hz) as our reference, the first resonance is raised in frequency and the second is lowered in frequency. Thus the spectrum envelope of [a] has two resonance peaks close together in the middle-frequency range between the first and second resonances of the neutral vowel.

F-Patterns

The relationship between vocal-tract shape and the frequencies of the resonance that arises from that shape is complex. Speech scientists have constructed idealized models of the vocal tract to simplify the calculation of that relationship.[12] Certain of those models are relevant to questions concerning the dimensionality of sound color and will be considered in more detail in the next section of this chapter. Here it is appropriate to examine the general effects on the spectrum envelope of

---

[12]The vocal-tract configuration of the vowel [aa], for example, has been approximated with a short, small-diameter tube section representing the throat connected to a tube of a larger diameter representing the relatively open front portion of the mouth. This is known as a *compound tube* model (Fant 1960).

various configurations of resonance frequencies. These configurations have been called by Fant (1960) *F-patterns*. An F-pattern for the vowel [a], for example, is 800, 1100, 2500, 3500, etc. The "etc." illustrates the utility of the term. When Fant says F-pattern, he is speaking of *all* the resonances that arise in a resonating tube, including the infinite number of higher formants, which are important more for their effect on the overall amplitude of the spectrum envelope than as peaks in the spectrum envelope at particular frequencies.

F-Patterns and Spectrum Envelopes

Fant (1960) has plotted, in a particularly revealing way, the spectrum envelopes of a systematic compendium of F-patterns. These drawings, reproduced in Figure 10, illustrate the broad range of spectrum-envelope shapes that are characteristic of the vowels. But more than that, they emphasize the effects of different F-patterns on the amplitude of the spectrum envelope in regions away from the resonance peaks.

Taking the neutral F-pattern as a reference, Figure 10 shows us that, as a general rule, when two resonances are close together in frequency the amplitude of the spectrum envelope at and above those resonances is raised in comparison to the reference. Conversely, when any two resonances are more widely separated than in the neutral F-pattern, the amplitude of the spectrum envelope between the two is lowered, as are the amplitudes of all the higher formants. Thus the low F_1 and F_2 in the F-pattern of [u] leaves a broad gap between F_2 and F_3, resulting in lower amplitudes of the third and higher formants. In the vowel [i], the close spacing of F_2 and F_3 raises the spectrum envelope in the F_2–F_3 region, but F_2 itself is low in amplitude because of the wide spacing between F_1 and F_2. The F-pattern of [a], on the other hand, has no large gap between formants, so the double peaks of F_1–F_2 are high in amplitude themselves and they cause the rest of the spectrum envelope to be raised in amplitude as well.

Singers will not find these observations surprising. They are well aware of the difficulty in projecting the vowels [i] and [u], a consequence of their relatively low-amplitude spectrum envelopes. The [a]

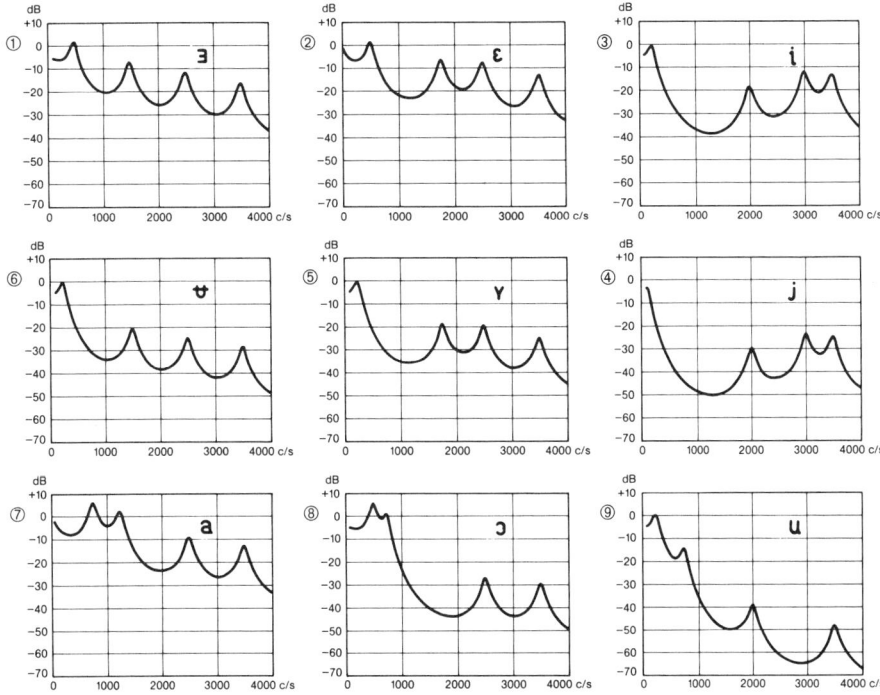

FIGURE 10: *Spectrum envelopes of some representative vowels (adapted from Fant 1960).*

vowel, on the other hand, is a singer's favorite, because its relatively high-amplitude spectrum envelope tends to reinforce the vocal source even away from the formant regions.

Sound Color and Resonances: Rule 1

Preliminary Rule 1′ associated sound color with the filter in weakly coupled, source/filter systems. Rule 1″ was more specific, isolating the spectrum envelope as the correlate of sound color. We now see that only the frequencies and bandwidths of the resonances in the filter need be specified. The spectrum envelope is determined in its entirety by the F-pattern, so the most efficient and revealing rule about sound-color invariance will be stated in terms of the F-pattern. Rule 1 is the answer, in final form, to the first fundamental question of this study.

Rule 1: To keep the color of a sound constant, keep the resonance frequencies (F_1, F_2, F_3, etc.) and bandwidths (B_1, B_2, B_3, etc.) of the filter constant.

Implied is the possibility of changing other aspects of the sound—its pitch, its loudness, and aspects of its timbre that do not have to do with color. Just so long as the F-pattern stays invariant—and, by implication, the filter and the spectrum envelope are also fixed—the sound color, according to Rule 1, will stay invariant.

Rule 1 is a hypothesis, first of all, and it is also a kind of definition of sound color. But most of all it is a psychoacoustic assertion that is susceptible to operational testing with appropriate psychological measurement techniques. In other words, it asserts that if we ask people—in some highly controlled and formal way—something about the color of a given sound, their responses will be consistent with Rule 1. The rule is not circular. Like many psychoacoustic relations, it is a claim that connects certain invariances in the acoustic and the perceptual realms, and it is testable just as postulated relations between loudness and intensity or pitch and frequency have been tested.

Limitations in the Scope of Rule 1

Rule 1 is meaningful (perhaps *practical* is the better word) only in weakly coupled systems. When the filter is connected to the source in such a way that the source is not independently controllable, we can hold the filter constant but we cannot at the same time change such other aspects of the sound as its pitch.

A potentially more far-reaching limitation of Rule 1 that must be recognized has to do with the effects of filter systems that are different in fundamental ways from the ones considered here. One such system is activated during nasal consonants and nasalized vowels. In these sounds the flap that closes off the nasal passages during the utterance of vowels—the fleshy uvula at the back of the soft palate—is lowered, opening the nasal cavities to excitation by the source. Speech scientists model such sounds with tubes that have two branches, one for the mouth and one for the nose. *Antiresonances* arising in the shunt paths of these more complex systems produce in the resultant spectrum envelopes V-shaped valleys and raised higher frequencies. Rules 1′ and 1″ are more general than Rule 1, because those preliminary rules can be construed as applying to any filter system, including branching ones with antiresonances in their spectrum envelopes. Strictly speaking, Rule 1 must be applied only to systems without antiresonances. The perceptual effects of antiresonances are secondary in importance to those of resonances, however. Even very complex filter systems with antiresonances have spectrum envelopes that resemble those of resonance-only systems. As a practical matter, Rule 1 may be applied even to filters that depart quite far from the "no antiresonance" stipulation.

Undoubtedly there are other limitations to the applicability of Rule 1. For example, there are upper and lower bounds to the lengths of resonating tubes, and hence of resonance frequencies, beyond which the rule will be invalid. Specifying these bounds is an empirical question, though, not a matter of principle. The rule can of course be applied to systems other than the adult male vocal tract. Even though the theory of sound color is inspired by the remarkably flexible mechanism of the human voice, it is more than a theory of vowel color; given certain restrictions, it is a theory about the perception of sound in general.

Another kind of limitation has to do, not with the filter, but with the form of the energy that excites the filter. This issue requires a brief side excursion into an examination of the characteristics of various sources.

The Adequate Source

By its very nature the source in the source/filter system is largely unconstrained. It can be irregular and uncontrolled, or it can be relatively simple.[13] The only question to concern us is whether the source spectrum has energy spread over a range of frequencies broad enough to

---

[13] Huggins (1952) has written an eloquent phenomenological description of the contrasting characteristics of source and filter. His thinking has had considerable influence on the present study.

excite at least the first two or three resonances in the filter. Let us call sources that have such a spread of energy *adequate*, and those without a sufficient spread *inadequate*. Guidelines are needed for predicting when a source will be adequate. Let us begin by examining certain very regular sources.

Ideal Sources

Nearly all waveforms that are easily specified mathematically, and whose spectra are easily derived, cannot exist in nature. They are known as ideal signals. Even though we never encounter precisely those sounds, we often hear sounds that are very close to them. And other sounds can sometimes be understood by interpolation between two ideal sounds.

The Sinusoidal Source

The sinusoidal source—energy at a single frequency—is clearly not adequate excitation of a filter. We can measure the output of the filter only at the frequency of the source; everywhere else it is zero. A number of sine waves at different frequencies or a sinusoid sweeping across the audible spectrum, on the other hand, can tell us a great deal about the filter. In fact, one of the best ways to measure the spectrum envelope of a filter is to excite it with a sweep tone and look at the output. This is the most common way of deriving the frequency response of any system; to all intents and purposes, frequency response and spectrum envelope are synonymous.

The Impulse

RECORDED EXAMPLE 2A

So far we have said very little about the effects of a filter system on the waveform of the source. It is often useful to view a sound in temporal detail, particularly when the bandwidths of the filter resonances are relatively narrow. The source used for this purpose is the *zero-width* pulse. The spectrum of this source has its energy spread over all frequencies, but at infinitesimal amplitude—the opposite case to that of the sinusoid. The spectrum of the filter's response to the pulse is also infinitesimal and therefore not very interesting. But the waveform is revealing. The response of a single resonance, for example, is a damped sinusoid (Figure 11). We can easily measure the resonance frequency—it is the reciprocal of the period of the damped sinusoid—and the bandwidth can be estimated from the rate of damping.

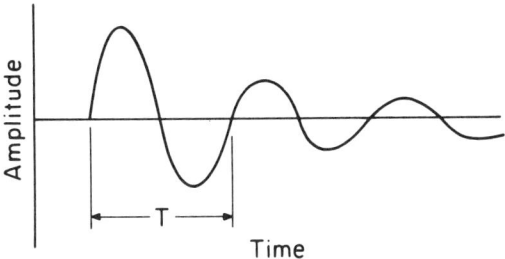

FIGURE 11: *Waveform of a simple resonator excited by a "zero-width" pulse. The frequency of the resonance is the reciprocal of the period, T.*

This way of looking at a filter casts some light on the example of the house renovator. The fingernail tap is a rough approximation to the ideal impulse, and the "pitch" that results is the ringing of the damped sinusoid in the impulse response of the wall. Under many circumstances a single impulse—at least the physically realizable, non-zero-but-very-narrow-width version thereof—can be an adequate source, as the case of the plaster-dusted wall tapper confirms.

The Pulse Train

A train of ideal pulses is similar in some ways to the vocal source in vowels. As in all complex, periodic signals, the components of this source are at frequencies that are integral multiples of the fundamental, i.e., the repetition rate of the pulse train. Adequacy depends on the fundamental frequency. When that frequency is low enough—and therefore has components that are relatively closely spaced—a good sampling of the spectrum envelope appears in the output of the filter. At higher frequencies, the pulse train is less effective, because the com-

RECORDED
EXAMPLE 2B

ponents are too widely separated in frequency. Just how high in frequency a pulse train can be while still making it possible for the ear to detect the sound color of filters with given F-patterns must be determined empirically. In theory, it is possible to determine resonance frequencies with remarkably widely spaced components.

Noise

RECORDED EXAMPLE 2C

Noise—a non-periodic signal caused by some sort of random process—is usually an adequate source and one that is very good at sampling the spectrum envelope of a filter system. The only exceptions are noises limited to high frequencies. Here, as in the case of the pulse-train source, determination of adequacy is an empirical question. Whispered speech is an example of noise as a source.

Guidelines for Adequacy in Sources

Consideration of these ideal sources leads us to conclude that for a listener to hear the color arising from a filter system, the source must have some amount of energy distributed across the frequency regions in which the filter resonances lie. A number of common sources tend to have at least some such energy distribution. Frequency modulation of a pulse train—as in a vibrato—spreads the energy, causing it to sample the spectrum envelope more broadly than the unmodulated signal and making an otherwise unacceptably high-frequency source adequate. Various kinds of irregularly timed signals have more-or-less noise-like spectra and are therefore likely to be adequate. Very regular sources, on the other hand, tend to be less adequate at exciting the filter resonances. Little more can be said about source adequacy. Generally, the issue should be settled empirically.

Distinguishing Source from Filter

Although most musical instruments are strongly coupled systems, in which the source and filter characteristics appear to be indistinguishable, many have multiple resonance systems, among which are some that are fixed in frequency—the plate resonances in the violin, for example. We would like to identify some cues in the sounds coming from such systems that would permit a listener to separate the effects of the fixed, weakly coupled resonance subsystems from the strongly coupled, pitch-determining subsystems. This is a particular case of a more general problem: On what basis can a listener distinguish the characteristics of a sound that are attributable to the source from those attributable to the filter? Evidence that pertains to the matter of how the auditory system actually makes the distinction will be reviewed in Chapters Three, Four, and Five. Here we can ask how, in principle, it can be made.

If we have knowledge of the source characteristics—or, as in the case of the wall tapper, we are in complete control of the source—we could simply perform, in the auditory processing mechanism of our brains, a comparison between our memory of the source and our present perception of the filtered version. The difference we would then attribute to the effects of the filter. Musical instruments with fixed resonance subsystems are difficult cases because we have in effect two filters, one that drives the source and another that is independent. One way of attacking this dilemma is to recognize the close association of the strongly coupled filter to the source by treating that whole system as if it were the source. Only the second, independent system is then treated as a filter. The experienced listener to instrumental music presumably knows a lot about the fixed filter subsystem of a particular instrument. That knowledge may be put to use in distinguishing two contrapuntal lines played on different instruments.

Two long-standing psychological theories of timbre suggest an acoustic basis for these kinds of distinctions (see Boring 1942, 368). The *relative pitch* theory holds that timbre is a function of the relative intensities of the partials in a complex tone—the equivalent of Rule X.[14] The *fixed pitch* theory is, in effect, Rule 1″. It says that, to keep timbre constant, a particular amplitude pattern in the components of a complex tone should *not* in general be held constant when the pitch

---

[14] Helmholtz is often said to have held this view of timbre, but he was at best ambivalent about it (e.g., Helmholtz [1877] 1954, 102–104, 124).

is changed. Rather, the component amplitudes of the signal with the changed pitch should be adjusted so that at particular frequencies the amplitudes are the same as they were in the original signal. We can see the effects of these two theories by comparing the spectra and waveforms of transformations that follow from each of them (see Figure 12). One way a listener can distinguish source effects from filter effects is to listen for a change of pitch and see which aspects change with the pitch change—the relative-pitch paradigm—and which stay the same—the fixed-pitch paradigm. The first of these would be associated with the source, the second with the filter.

A broad range of signals—nearly all sounds in nature, for example—are potential sources. This alone suggests the considerable musical potential of sound color. Simply by changing the source that is exciting a pattern of fixed-filter-system settings, a musical structure built on sound color can be transformed in essentially unlimited ways.

THE DIMENSIONS OF SOUND COLOR

Let us consider once more the analogy with visual color. Visual color results from light having a preponderance of energy at certain wavelengths. We see red when the light reaching our eyes has a wavelength of about 700 nanometers (nm). We see yellow with light at 575 nm, green at 510 nm, and blue at 460 nm. Our experience with visual colors tells us that there is a special relationship between red and green on the one hand and yellow and blue on the other. Each of these pairs, when mixed together in equal amounts, results in the "non-color" gray. These pairs, and all other pairs whose mixture results in gray, are called complementary colors.[15] The most common way to represent these color mixtures is to arrange the various colors in a "color circle," with complementary colors on opposite ends of lines drawn through the center of the circle and gray in the middle. Many other phenomena

---

[15] These facts about color mixture apply to so-called spectral colors—produced by passing white light through a prism—not to pigments, for which a different set of mixing rules apply.

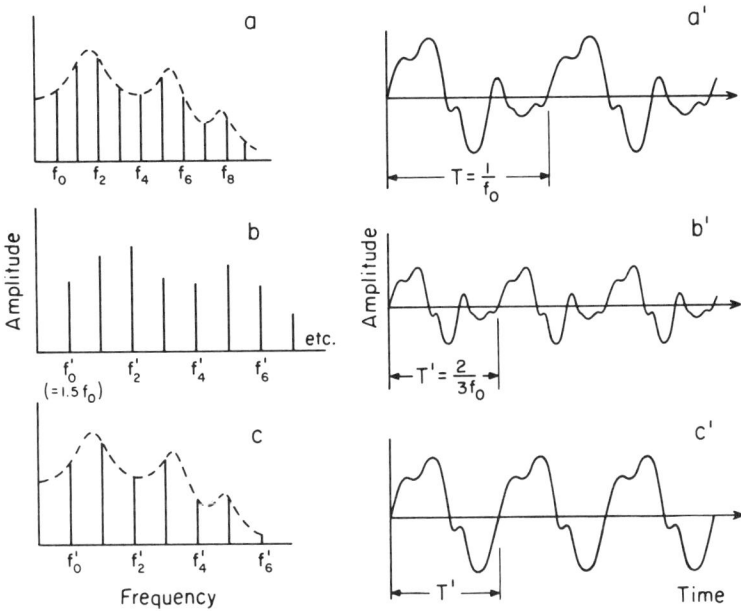

FIGURE 12: *Relative-pitch versus fixed-pitch transformations. (a) The spectrum of a complex sound; (a'), its waveform. (b) and (b') The relative-pitch transformation when the fundamental frequency is raised by a fifth ($F_0 * 1.5$). (c) and (c') The fixed-pitch transformation.*

of color vision can also be represented by this circle.[16] The color circle, then, implies that our color sense—the perceptual "space" of visual color—is multidimensional. Although the physical continuum, measured according to the wavelength of light, is a single dimension, the *psychological* representation of that continuum requires at least two dimensions in order to specify the color circle.

Many of our senses have this multidimensional character. Sometimes, as in taste, there are fairly direct relationships between physical quantities and psychological dimensions—acidity and sourness, for example. In other cases, as in color vision, the relationships between the physical and psychological continua are more complex. Rule 1

---

[16] Lindsey and Norman (1977) provide an excellent, readable discussion of color vision.

implies that sound is also multidimensional in that sound color and other features of the sound, including pitch, are independent of each other.[17] The elaborate classification system of Schaeffer (1968) suggests that a description of timbre alone requires many dimensions. It is not surprising, then, to find that sound color itself is made up of more than one aspect or dimension.

The second fundamental question posed in Chapter One simply assumes that sound color has different aspects; that, in other words, it is a multidimensional attribute of sound. In fact, not all uses of sound color support that assumption. For example, the methods of subtractive synthesis in electronic music only occasionally provide justification for attributing more than a single dimension to sound color. Typically only a single filter is used, either fixed in frequency or sweeping over a range of frequencies. The resulting sound colors are often described as though they varied along a single continuum, say, from bright to dull.[18] Quite a different picture is presented by speech. Speech sounds are clearly multidimensional. The vowels alone provide considerable reason to regard sound color as a multidimensional psychological attribute. The character of the vowel-like F-patterns (such as those in Figure 10) suggests strongly that vowels cannot be characterized reasonably by values of a single variable. That suggestion is supported if we reflect in a general way about how we perceive vowel sounds.

The vowel sequence [u ("boot"), o ("boat"), ɔ ("bought")] seems unidimensional, as does the sequence [u, ø (German ö), i ("beet")]. The changes are different in the two cases, but they are equally natural and orderly. If we wanted to fit both vowel sequences into a single dimension, we might combine them into a single continuum: [i, ø, u, o, ɔ]. But then how shall we treat the sequences [i, e ("bait"), æ ("bat")] and [æ, a ("hot"), ɔ]? These seem as naturally grouped as the first two, but now we have come round full circle; a single-dimen-

---

[17] Lindsey and Norman (1977) also provide a good general introduction to the dimensions of hearing.

[18] Some musical examples of such treatment of filters will be discussed in Chapter Six.

sional representation is ruled out. Moreover, our circle does not include the neutral vowel or the various "short" vowels, such as those in the words *bit*, *bet*, *but*, and *put*. Still closer observation[19] shows something of the articulatory basis for these perceptual groupings. In pronouncing the vowel sequence [u o ɔ], we begin with the lips very close together and then open the mouth successively wider. A very similar series of motions is involved in going from [i] to [e] to [æ]. A more subtle, but still detectable, comparison can be made with respect to the degree of lip protrusion—called by phoneticians *rounding*—in the sequences [u ø i] and [ɔ a æ].

These informal observations suggest that we need at least two dimensions, and possibly more, to account for the perceptual categories of vowels. Since sound color is conceived as having a large domain that includes vowel color, we can conclude that sound color too requires multiple dimensions that should, in some sense at least, include the dimensions of vowel color. What are those dimensions? We can begin to answer that question, in principle, by drawing on the distinctive features of vowels as posited in linguistic theory.

Distinctive Features and Sound-Color Dimensions

The *distinctive feature theory*, as proposed by Jakobson, Fant, and Halle (1951) and revised and refined by Chomsky and Halle (1968), codified certain long-standing observations of phoneticians by hypothesizing that the many sounds of speech can be placed in categories based on the presence or absence of certain distinctive features. Whether the mouth is open, whether there is a narrowing of the vocal tract at a particular place, whether a consonant is aspirated—properties such as these make up the features that characterize and distinguish the phonetic content of a language. The theory is powerful because it can be applied, with only slight modifications, to all human languages throughout the world.

Three of the distinctive features of vowels stand out as candidates

---

[19] Actually pronouncing the examples—ideally, while looking in a mirror—will aid in following the argument here.

from which to derive dimensions of sound color.[20] The first feature is called *compactness*. The vowels /i u/ are non-compact; the vowels /æ ɔ/ are compact. *Acuteness* distinguishes between vowels like /u ɔ/ and vowels like /i æ/; the former pair are non-acute, the latter pair are acute. The third feature, *laxness*, makes a different sort of distinction; it separates the long vowels from the short vowels. The vowels in *beet*, *bat*, *bought*, and *boot* are non-lax, whereas those in *bit*, *bet*, and *put* are lax. To distinguish all the vowels in English certain additional features are required, but these three, compactness, acuteness, and laxness, account most satisfactorily for what may be considered extreme configurations of the vocal tract during the pronunciation of vowels.

The sound-color dimensions that correspond to these three features we shall call OPENNESS, ACUTENESS, and LAXNESS.[21] A fourth dimension, SMALLNESS, has no corresponding vowel feature, and it will be discussed separately below. The name of the first dimension has been changed from the rather misleading *compactness* to the more descriptive OPENNESS—a reflection of our observations that the compact vowels /æ a ɔ/ are spoken with a wide-open mouth in contrast to the non-compact /i ø u/, spoken with a narrow mouth opening.[22] We are able, with this choice, to avoid calling sound colors such as [aa] and [ae] "compact" and sound colors like [uu] and [oe] "non-compact"—a perceptually counterintuitive terminology.[23] The names of the other two dimensions are borrowed directly from the original statement of the theory by Jakobson et al. (1951).

Unlike the linguistic distinctive features from which they are derived, the sound color dimensions are intended to represent perceptual

---

[20] The names of the features discussed here are taken from Jakobson et al. (1951). The revisions by Chomsky and Halle were justified on linguistic grounds (1968, 306) and need not concern us here.

[21] The dimensions of sound color will be distinguished typographically in this study by printing their names in small capitals.

[22] Phoneticians themselves often speak of "open" versus "narrowed" or "closed" vocal-tract configurations in spite of certain linguistic objections to this usage. Since sound color is explicitly pre-linguistic, the term that is descriptive of the tube shape is preferred.

[23] I am grateful to Robert Cogan for pointing out this terminological problem.

continuities, not binary distinctions.[24] We would therefore like to specify sound-color dimensions as continuous functions of physically measurable quantities. Measures of tube shapes would be one such set of quantities. The F-pattern, which determines the spectrum envelope and represents a filter's effect on a sound, is a more general level of description that is not tied to a specific filter type. The dimensions of sound color will therefore be specified as functions of the filter resonance frequencies; how tube shapes correlate with the dimensions will be discussed only by way of illustration.

One further simplification is needed. A long tradition in the analysis of vowels[25] has demonstrated the strong dependence of vowel quality on the frequencies of the first two formants alone. Some of the evidence for this conclusion is reviewed in Chapters Four and Five, and Figure 10 suggests, from mathematical considerations alone, how powerful the first two resonance frequencies' effect on the spectrum envelope is in comparison to that of the higher resonances. It is known that the higher resonances must be adjusted from their neutral positions in certain vowels (notably /i/ and /ø/; see Fant 1959). However, these adjustments are not ordinarily independent; they are almost always associated with particular values of F_1 and F_2.

Taking our cue from research on the vowels, we assume, for the purpose of specifying the sound-color dimensions, that sound color is primarily a function of the frequencies of the first two resonances. This is not to say that the frequencies of the higher resonances have no effect

---

[24] The difference between features and dimensions is not equivalent to the contrast between the classificatory and and the phonetic functions of features (Chomsky and Halle 1968, 65). Features, in the latter sense, can take on multiple values that correspond to neurological instructions to the speech mechanism, whereas the sound-color dimensions may have values that are not realizable by that mechanism. Nevertheless, we know that the multi-valued phonetic function of the features is expressed ultimately by continuous motions of the articulatory mechanisms, and our auditory systems deal with the acoustic result of these motions. This conversion of binary features into continuities in speech production suggests and supports treatment of the dimensions in the theory of sound color as continuous.

[25] See, for example, Peterson and Barney (1952) and Chiba and Kajiyama (1941).

on sound color, but experience with the vowels suggests that those effects are secondary. It makes sense, therefore, to attempt to specify the most salient dimensions of sound color in terms of the frequencies of the first two resonances alone.

Equal-Value Contours: Specification of the Dimensions

The second fundamental question asked in Chapter One dictates the form specifications of the dimensions will take. Again we want to know how to hold one dimension of color constant in the face of variations in the other dimensions. For each dimension, what is needed is a family of loci, defined in terms of the first two formant frequencies, on each of which the applicable dimension of color does not vary. In other words, we need to specify contours—like equal-altitude gradients on a map—of equal OPENNESS, equal ACUTENESS, equal LAXNESS, and equal SMALLNESS.

Figure 13 presents examples of these contours plotted on the F_1–F_2 plane. The x-axis in these graphs is the frequency of the first resonance; the y-axis, the frequency of the second resonance. The lower right-hand corner of this plane contains an area (marked out in the figure) in which the second resonance would be lower than the first, so this area is forbidden.[26] The F-patterns of Figure 10 may be located on the graph, as may those of all the vowels. These contours represent, as closely as our present knowledge permits us to specify them, the dimensions of sound color.

OPENNESS

As suggested above, the first dimension is named for the tube shape with which it is correlated. The approximate acoustic correlate of OPENNESS is the frequency of the first resonance. It follows that the equal-OPENNESS contours on the F_1–F_2 plane are nearly vertical lines.

---

[26] There is nothing mysterious about this crossed-formant area: points falling within it are reflected out of the region by the convention that the lowest frequency resonance is always called F_1.

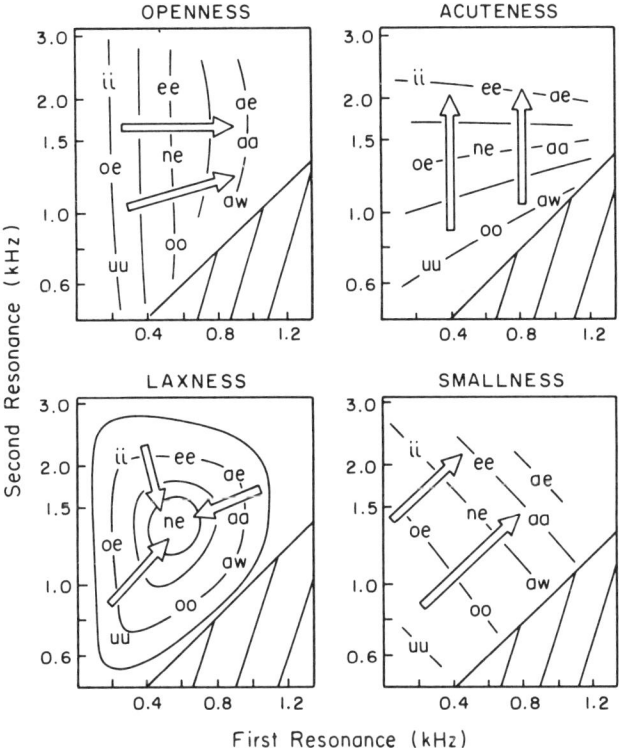

FIGURE 13: *Contours of equal* OPENNESS, *equal* ACUTENESS, *equal* LAXNESS, *and equal* SMALLNESS. *The equal-value contours are plotted as functions of the frequencies of the first two resonances, F1 and F2. Arrows indicate direction of increasing value. (Adapted from Slawson 1981.)*

Some empirical considerations—the locus of the vowel /a/, in particular—suggest that the high OPENNESS contours may be bowed outward slightly. The colors [uu oe ii] are all equal and low in OPENNESS, whereas the group [aw aa ae] are equal and high in OPENNESS.

ACUTENESS

Reflecting its everyday connotation of "high" or "bright" sound, the second dimension, ACUTENESS, increases with increasing frequency of the second resonance. Its equal-value contours are thus approximately horizontal lines on the F_1–F_2 plane. The exception is at high values of the first resonance, where the low ACUTENESS contours slant up slightly. The equal-ACUTENESS contours can be read as indicating, for example, that the sound colors [uu oo aw] all have equal and low ACUTENESS and that [ii ee ae] have equal and high ACUTENESS.

LAXNESS

The maximally LAX point is an ideal in several senses of the word. This is the F_1–F_2 locus of the uniform tube equal in length to the average adult-male vocal tract. These also are the formant values that would arise, in theory, from the vocal mechanism in the position to which it is automatically brought just before beginning to speak (Chomsky and Halle 1968). As is suggested by its name, LAXNESS is said to correspond to a relatively relaxed state of the articulatory musculature (although this notion has been criticized: Lass 1976). The equal-LAXNESS contours are closed curves on the F_1–F_2 plane centered on the maximally LAX point. The outer curve represents low LAXNESS and, as has been suggested above, connects the loci of all the long vowels. The short vowels fall on a medium LAXNESS contour.

SMALLNESS, an Additional Dimension

One further dimension of sound color, called SMALLNESS, has no parallel in the features of vowels, but it can be justified by its rough correspondence to the overall length of an acoustic tube or the overall size of other sorts of resonators. The name of the dimension comes from that approximate correspondence. The equal-SMALLNESS contours in the F_1–F_2 plane are diagonal lines that link the colors [ii ne aw] at medium SMALLNESS, [oe oo] at medium-low SMALLNESS, and [ee aa] at medium-high SMALLNESS. The relationship to size depends on the very general rule that large (or long) objects tend to resonate at low frequencies and small (or short) objects tend to resonate at high frequencies. Thus [uu], having the lowest values of both F_1 and F_2, is very low in SMALLNESS, and [ae], with high F_1 and F_2, has a very high SMALLNESS value. The colors that fall between these extremes have SMALLNESS values that vary approximately with the mean of F_1 and F_2.

Sound-Color Space: Rule 2

We are now ready to state Rule 2, an answer to the second fundamental question from Chapter One.

> *Rule 2:* Sound color has the dimensions of OPENNESS, ACUTENESS, LAXNESS, and SMALLNESS. These dimensions are defined by the equal-value contours shown in Figure 13.[27] To hold sound color constant with respect to one dimension, change the values of F1 and F2 in such a way as to remain on one of that dimension's equal-value contours.

Rule 2, including Figure 13, specifies the relationships between perceptual aspects of sound color (the dimensions) and acoustic quantities (the frequencies of the first two resonances). By specifying relationships among the dimensions themselves, it also provides a structure for the entirely perceptual "space" of sound color. For example, we can infer from Rule 2 that the maximally LAX color area corresponds to medium values of OPENNESS, ACUTENESS, and SMALLNESS, whereas all extreme values, high or low, of the latter three dimensions are of minimal LAXNESS. Sound colors with high ACUTENESS and low OPENNESS are of medium SMALLNESS, as are colors with low ACUTENESS and high OPENNESS. The interrelationships of the dimensions will be discussed in greater detail in Chapter Seven; here we can observe that they are to some extent independent of the psychoacoustic specifications of Rule 2. Even if research were to find that the equal-value contours of Figure 13 are somewhat misplaced, the structure of the sound-color space—i.e., the interactions of the dimensions—might not be affected.

As in the statement of Rule 1, we are left with a number of theoretical issues that require resolution. The domain and status of the specifications of the sound-color dimensions remain to be clarified, and the statement of Rule 2 raises several issues that have been of concern to

---

[27] This graphic specification of the equal-value contours is not exact, but the psychoacoustic evidence, to be reviewed in Chapter Four, is not yet of sufficient quality to attempt a mathematical specification.

psychoacousticians in other contexts. Let us begin with a fairly straightforward convention for talking about values of sound color with respect to each of the dimensions.

Measurement in Sound-Color Space

We shall treat the maximally LAX point as a zero reference for the other three dimensions. The equal-value contour that passes through that point will be called the *zeroth contour*. Contours of OPENNESS, ACUTENESS, or SMALLNESS that are higher than the zeroth contour will be said to have positive values; those lower in OPENNESS, ACUTENESS, or SMALLNESS, negative values (see Figure 14). Thus the color [ae] is positive in OPENNESS, ACUTENESS, and SMALLNESS, whereas [aw] is positive in OPENNESS, negative in ACUTENESS, and has zero SMALLNESS. The dimension of LAXNESS has a real, not an arbitrary, point of discontinuity—the neutral position. It is less useful to have a sign convention for that dimension. When such is called for, we shall assign a value of zero to the point of maximum LAXNESS and assign negative values of increasing magnitude to the surrounding equal-LAXNESS contours as they decrease in LAXNESS. It follows that there are no positive LAXNESS values.

The sign convention just introduced attributes no special significance to the zeroth contours of OPENNESS, ACUTENESS, and SMALLNESS, nor does it impose a particular metric on the sound-color space. Thus, strictly speaking, Rule 2 should be considered qualitative, not quantitative.[28] It is claimed that the dimensions are continuities; thus, the contours in Figure 13 are only samples drawn from an infinite family of such contours. The general shapes of the equal-value contours are claimed to be like those plotted in Figure 13, but the theory asserts no precise psychoacoustic relationships. However, some empirical studies, reviewed in Chapters Four and Five, suggest that the differences between the equal-value contours as drawn in Figure 13 may in fact represent approximately equal differences in sound color.

[28] It can be suggested that the psychological continua along the dimensions of ACUTENESS, OPENNESS, and SMALLNESS are probably what S. S. Stevens (1951) would call "metathetic," requiring no more than an "interval scale." The LAXNESS scale, since it has a natural zero point, may be closer to a "prothetic" continuum.

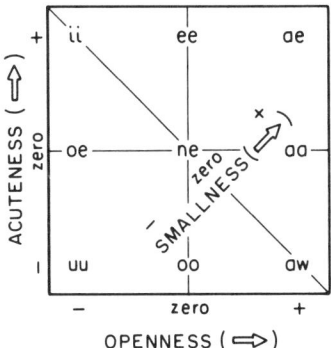

FIGURE 14: *A sign convention for sound-color space. Equal-value contours that pass through the neutral position are called zeroth contours.*

For illustrative purposes in the following sections, specific values of a sound color with respect to the dimensions will sometimes be assumed, and appropriate arithmetic operations on those values will be performed. Even though the psychoacoustic determination of the scales is left open to future empirical study, the theoretical claims made by Rule 2 are substantive and falsifiable. If appropriate, thorough, and sensitive psychoacoustic studies found invariances in sound color whose loci were markedly different from those of Figure 13, the disconfirmed portion of the theory would have to be revised.

Independent Constancy versus Independent Variation

Rule 2 and the question it purports to answer are expressed in terms of sound-color invariance. To hold a color constant with respect to one dimension while letting it vary with respect to other dimensions—that is, independent constancy—is straightforward enough; to vary a sound color along one dimension while holding it constant with respect to all other dimensions—that is, independent variation—is quite another issue. Because OPENNESS and ACUTENESS are approximately orthogonal to one another, it is possible to vary color with respect to one of those dimensions without changing it with respect to the other. However, such variation will almost always result in changes in the values of LAXNESS and SMALLNESS. Moreover, except in certain very specific cases, changes in LAXNESS will result in changing values of

OPENNESS, ACUTENESS, and SMALLNESS. Independent constancy is a weaker condition than independent variation, but it suffices for the description and measurement of many psychological attributes. Such attributes, or dimensions, of perception arise out of our ability to focus our attention on one of a number of different, non-orthogonal but no less psychologically real aspects of a sensory stimulus (cf. Stevens, Guirao, and Slawson 1965).

Alternative Formulations of the Dimensions

The formulation of the sound-color dimensions presented here is based largely on rational or principled grounds. An alternative approach to the topic might be to collect a wide variety of sounds and then perform one of the multidimensional scaling procedures in an attempt to derive the dimensions entirely on an empirical basis. Some researchers have made such attempts and their results will be reviewed in Chapter Four. We cannot be entirely comfortable, however, with generalizations about sound color based solely on such determinations. So-called generalizations derived empirically are seldom very general, and they are limited, in nearly every case, to only the most salient characteristics that emerge from the variability in a particular series of experiments. Nor is experimental "noise" the only drawback. Every procedure, no matter how sophisticated, introduces biases that tend to distort the results in ways that may not be detectable and for which there are no predicted values for comparison.

The approach taken in this study is to postulate a set of dimensions on rational grounds and then to test that theoretical formulation experimentally. This approach has its weaknesses as well—for example, the theory can be mistaken, it can turn out to be untestable, or its domain may be shown to be very limited—but it has the critical advantage of permitting the development of a logical structure that can be applied to a number of problems, including musical ones.

The Sound-Color Dimensions as Pre-Speech

As Chomsky and Halle argue in their elaborate and closely reasoned study (1968), the phonetic features can be construed to be substantive linguistic universals. They must, therefore, have a biological basis.[29] It would be naive to conclude, however, that because of this link to biology, the perception of all sound should be treated as governed by phonetic theory. Modern work in ethology has demonstrated that innate behavior is "released" by a specific small range of stimuli. For the full-blown process of categorical distinctions called for by the distinctive-feature theory to be brought into play, human beings may well have to be *predisposed* for spoken communication; they may, in other words, have to be listening in a "speech mode."

On the other hand, we have reason to expect that sensory systems—presumably including the auditory system—must structure to a considerable degree *all* the information they transmit to higher brain centers.[30] In the auditory system, this kind of complex auditory processing would apply to all sounds and could thus be involved in the perception of sound color. Unfortunately, we have only limited clues from auditory physiology about what kind of processing actually takes place. Attempts in this area, which we shall review in Chapter Three, have met with a number of difficulties. What sorts of phenomena might we expect to find in theory? One way of answering this question is to reexamine the distinctive features. If we can identify certain primitive features of speech that serve some pre-speech function, we have reason to consider their inclusion among the features of sound in general and of sound color in particular.

Rule 2 can be interpreted as a claim that the dimensions of OPENNESS, ACUTENESS, SMALLNESS, and LAXNESS are fundamental biological features that are part of the auditory processing of all sounds. The color of a sound is determined by its value on each of the dimensions, and its phonetic category—in the speech mode—may be determined by which side of a critical point on the dimensions its sound color lies. K. N. Stevens (1972) has presented the interesting argument that certain vocal-tract configurations, and their corresponding F-patterns, are "polar" and thus susceptible to easy categorization and binary fea-

---

[29]This point is developed explicitly by Lenneberg (1967).
[30]This is most clearly indicated by studies of the physiology of the visual system, e.g., those of Hubel and Wiesel (1962).

ture extraction. The possibility that we can identify certain F-pattern extremes particularly easily may be very relevant to a discussion of sound color. In fact, the criteria Stevens uses to determine which configurations have critical characteristics can be applied, in part, to *any* compound tube. We can use these special configurations as clues to, and arguments for, continuous sound-color dimensions that connect these extremes in a pre-speech mode. Sound color, according to this formulation, is a primitive attribute of sound—possibly also perceived by animals besides man—that provides a kind of substratum for the perception of speech, and some of the characteristics of which we can infer from the categories of speech.

We can summarize this rather intricate argument in terms of a suggested scenario for the origins of speech in the prehistoric past. Let us assume that the auditory system of primitive man, like that of many other animals, was capable of analyzing the color of sounds along several dimensions. For one of many reasons—or perhaps several of them coinciding—man began to develop intricate control over his vocal mechanisms. This process was supported and reinforced by the built-in analyzing capability of his auditory system, which could be exploited to carry information to his brain about the positions and movements of his vocal tract. The stage was set for the development of the categories of speech and of the higher levels of language.

This outline is vastly oversimplified, and undoubtedly wrong in detail, but it emphasizes that considerable auditory capacity is a prerequisite for speech communication. It is natural to look to speech for clues to the character of that auditory capacity, an important portion of which, according to the present theory, consists of the analysis of sound color.

Normalization of the Sound-Color Space

Sound color was said, in the discussion above, to encompass a larger domain than vowel color in that it may be applicable to sounds arising in objects that are larger, smaller, or differently shaped than the adult male vocal tract. Our auditory systems may attempt to analyze color in all sounds according to the dimensions, as was suggested above. Alternatively, it may be that different kinds of resonators will require another system of dimensions. In the general case of an arbitrarily complex resonator, we have no rationale for altering the basic theory or adding new dimensions. What we can deal reasonably with are sounds that depart in certain very regular ways from the "normal" sound-color space—in particular, those that arise in tubes of different lengths.

Straightforward mathematical considerations (Fant 1960) lead to the conclusion that changing the length of a tube closed at one end while keeping the shape of the tube the same causes the formant frequency to be multiplied by a constant factor. It follows that such proportional changes in filters require that we extend or contract the sound-color space of Figure 13, keeping the equal-value contours in their relative positions. But Figure 13 deals only with the first two formants; changes in length also require shifts of the higher resonances by the same factor. We must turn to these higher-resonance changes for cues for the perception of such shifts. The reason for this can be seen most easily by following what happens to particular colors when such shifts take place.

For example, let us assume that the 17-centimeter tube is shortened by 25 percent. The neutral color in the center of the sound-color space would have its first resonance multiplied by a factor of $1 \div 0.75$ or 1.333. F_1 would change from 500 Hz to 667 Hz, F_2 from 1500 Hz to 2000 Hz. If we ask whether this shift has resulted in a change of sound color, we are faced with a dilemma. At first glance (see Figure 15), we would have to say yes. The F-pattern has changed from the neutral color to one that is less LAX, in the direction of ae, so according to Rule 1 the sound color is no longer invariant. On the other hand, if to accommodate the shorter tube we have stretched the F_1–F_2 plane with its equal-value contours intact, we have moved the [ae] color to a higher F_1–F_2 locus and the new neutral point is still in the middle of the *perceptual* sound-color space. Figure 15 illustrates these alternatives. There may be some kind of generalized change in the sound color, but, paradoxically, the sound-color values on each of the dimensions—newly configured—have not changed.

When we are listening to a sound, our only reasonable basis for deciding whether a change in an F_1–F_2 locus is a genuine color change or is the result of an overall shift in the space lies in the frequencies of the higher resonances. Suppose, first, that we hear no change in the

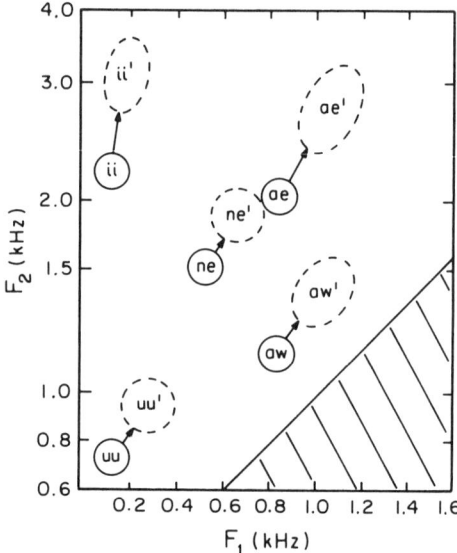

FIGURE 15: *Unshifted and shifted sound-color spaces. Shifted versions are indicated with primes and dotted areas.*

higher resonances. We must then conclude that there has been no change in the overall length of the tube, and that the F_1–F_2 shift must have been due to a perturbation away from the uniform cross-section of the neutral tube. Suppose, on the other hand, we hear an upward shift in the frequencies of the higher resonances along with an upward shift in F_1 and F_2. In this second case we have a complete F-pattern that is consistent with a shorter version of the uniform tube and we have reason to regard the color change as due to a global expansion of the sound-color space. The values of the shifted sound color with respect to each of the (shifted) dimensions will remain the same.

An argument of this sort has been used to explain the perception of children's speech and, to a lesser extent, that of women. The vocal tracts of women and children are shorter, on the average, than those of men and so produce higher, more widely spaced upper formants that are cues for what is called the *normalization* of their vowels.[31] We can

---

[31] Fant's review (1973a) of the effects on formant frequencies of the vocal tract dimensions that are typical of females and children points out certain changes in formant spacings aside from a proportional increase in frequency.

adopt that term to apply to the adjustments we make when we are listening to the colors of sounds in which the upper resonances of the F-pattern lead us to conclude that the sounds are produced in a tube longer or shorter than normal. When we normalize a sound color, we convert it back to the space specified in Figure 13. Or, in what amounts to the same thing, we adjust Figure 13 to the shifted sound.[32]

Because much of the burden of justifying the SMALLNESS dimension rests with the effects of differences in size of the resonator, there is a danger of confusing that dimension with the shifts that call for normalization. The equal-SMALLNESS contours, like those for the other dimensions, are defined in terms of F_1 and F_2 alone; they do not involve resonances above the second. Shifts due to specific changes in tube length, on the other hand, are cued entirely by the frequencies of the upper resonances. By hypothesis, then, SMALLNESS values, along with those of the other dimensions, are subject to normalization when the upper-resonance frequencies call for it.

Limitations of the Dimensions

Rule 2 and the normalization process have certain limitations that must be recognized. Unfortunately, few of these limitations can be discussed as matters of principle; their specification will have to depend on future empirical study.

Resolution of F-Pattern Analysis

Both Rule 1 and Rule 2 describe perceptual constancies in terms of invariances or specific changes in resonance frequencies. The question arises, how invariant or how specific a change? How large a deviation from a constant spectrum envelope or from an equal-value contour can

---

[32] It should be pointed out here that the choice of the average adult male vocal tract as "normal" is only a convention that follows from the relatively large amount of research that has been done on adult male speech. The theory is not affected in any significant way by starting with, for example, the F-patterns of an average adult female and normalizing "down" to derive adult male F-patterns.

we make and still retain the perceptual invariance? The few relevant experimental studies of this question will be reviewed in Chapter IV. Here we must simply acknowledge that the capability of the auditory system for analyzing the frequencies of resonances in a spectrum envelope is limited. We ought also to keep in mind that no single value for resolving power is likely to be valid. Musical context is probably only the most powerful of a large number of factors that strongly affect the size of the just-noticeable-difference in sound color.

The Domain of Color-Space Shifts

Changes in overall filter size account for only a limited range of shifts of the sound-color space. If, for example, a tube is very long, the resonance frequencies may fall too close together to be perceived as effective in determining sound color. Conversely, very short tubes may create resonances so far apart in frequency that sound colors become indistinguishable. However, we have no theoretical definition of the locations of these absolute limits to the sound-color space. Some measurements (to be reviewed in Chapter Five) suggest a range of formant shifts in vowels within which normalization may take place, but we have no information about where normalization may break down. Here too we must be concerned with context. A drastically shifted sequence of sound colors, which in juxtaposition with its unshifted version may sound quite unrelated, could be linked to the original by a musical context in which the shift is accomplished gradually over some span of time. There are undoubtedly other factors that affect the size of the sound-color space within which normalization is possible.

Sound Color in Aberrant Resonators

A vast collection of sound-producing objects are quite unlike the human vocal tract. Uniform tubes that are open at both ends, for example, have resonances at all multiples of the first-resonance frequency, not, as in the "quarter-wavelength" tube of the vocal tract, only at odd multiples of F_1. Plates, membranes, and bars have a vast variety of resonances depending in a complex way on the material makeup, the mass, the stiffness, and the shapes of those objects. A branch of physical acoustics has undertaken the analysis of the simplest of these resonance systems (e.g., Morse 1948). How does the concept of sound color, which is derived largely from a very specific type of resonator, apply to sounds produced in these other objects? There is no good theoretical answer to this question. The few experimental studies of the perception of musical instrument timbres (reviewed in Chapter Five) provide no convincing evidence for additional dimensions that might be added profitably to the four postulated in Rule 2. We can probably assign certain of the characteristics of sounds in resonators, particularly those that arise in high-Q devices, to other categories of musical timbre, such as Schaeffer's "allure." But it may be that sounds requiring dimensions other than those proposed here for their proper analysis should be included under the rubric of sound color. Study of such sounds will, however, have to be left for the future.

One rather common practice—that of imitating the sounds of musical instruments with the voice—suggests that sounds made in a variety of ways may be heard as if they were produced in a vocal tract–like resonator. Brian Fennelly (1967) has proposed a notation for electronic music that recognizes this connection of timbre in general to vocal color. We can imagine that the crowded spectrum of a musical instrument's sound might have a formant structure—broad maxima imposed by some sort of "super" spectrum envelope—that would form the acoustic basis for this suggestion. A few experimental studies that demonstrate such phenomena in the sounds of certain musical instruments will be reviewed in Chapter Five. The auditory system may well be predisposed to detect such spectral maxima, since the early development of speech in man's prehistory very likely involved small adaptation of the ear as well as the more morphologically obvious adaptations of the vocal tract. If the ear is in some sense biologically conditioned for the recognition of speech, then it may not be too farfetched to suggest that our auditory systems, on some level, may analyze sounds as if they were speech-like whether or not they are produced in a speech-like manner.

OPERATIONS ON SOUND COLOR

All the well-known ways that composers vary musical entities can be construed as musical *operations*. Viewed in this general way, musical operations can be seen as the key to coherence at all levels of structure and form. Without the appropriate operations of embellishment, the performance of a raga would be brief indeed, a set of variations would be impossible to compose, jazz would consist only of the tune repeated over and over. Transposition is a necessary feature of nearly all the forms of music written during the "common practice" period of concert music in the European West. In the atonal and twelve-tone music of the twentieth century, inversion and retrogression supplement a version of transposition as the basis for holding certain pitch properties invariant while varying others. These are only the best known of a rich palette of such operations with which pitch and, to a lesser extent, rhythm have been structured.

The fact that pitch operations were well known was, in effect, what Schoenberg ([1911] 1978) meant when he spoke of the "laws" for structuring pitch. And the lack of such specific operations in the realm of tone color led him, quite correctly, to characterize that area of composition as guided by "feelings" alone. The theory of sound color as presented to this point does not yet contain rules for transforming sound colors, but the stage has been set. Rule 1 gives us a basic acoustic correlate of sound color, and Rule 2 tells us how the sound-color space is organized. Although the dimensions of sound color do not generate a set of sound-color operations in any automatic sense, their relations to each other and to the point of maximal LAXNESS—the neutral position—suggest such operations.

General Characteristics of Operations

The independence of sound color is emphasized by certain characteristics that all sound-color operations share and that differ significantly from their pitch analogs. Sound-color operations, unlike those in the realm of pitch, are defined *with respect to one of the dimensions*. Thus, there are no sound-color transpositions or inversions per se but, rather, transpositions or inversions with respect to, e.g., ACUTENESS or OPENNESS. As a result, there are more fundamentally different sound-color operations available to a composer than there are pitch operations.[33] Because the sound-color space contrasts in certain formal ways with the pitch spaces, the operations too contrast formally. For example, in designing or analyzing a passage in atonal music, one must be concerned both with pitch classes and the intervals between them, and also with the actual pitches in their actual registers. But in the realm of sound color there is no octave and hence no distinction between an "interval space" and a "register space." The sound-color space may be enlarged or contracted from its normal configuration, but there seems to be little justification to appeal to a concept of "color classes." Once a color space has been normalized, all such classes would be treated as a single color; nothing comparable to pitch register need be considered.

In the statements of Rules 1 and 2, the connection of color and its dimensions with values of the filter resonances is made. Those rules are sound-color analogs to the well-established psychoacoustic relation between pitch and the log-frequency scale. Conversely, the definition of the sound-color operations—like the definitions of pitch inversion, embellishment, etc.—is made entirely within the perceptual or psychological side of the psychophysical dualism. That is to say, we assume that the dimensions are related *to each other* as stated in Rule 2 and illustrated in Figure 13. Then the *perceptual* space, defined by the dimensions, becomes the domain over which the operations are defined. Even if the dimensions were to be found to have acoustic correlates that are somewhat at variance with Figure 13, the operations might not need to be redefined. As long as the dimensions relate to each other essentially as indicated in that figure, the operation will stand.

Sound-Color Transpositions

Pitch transposition is perhaps the most perceptually compelling of the pitch operations. Although, as we shall see, sound-color transposition may not always be as "natural" as its pitch analog, the two share certain

---

[33] This point is made more specifically in Chapter Seven.

characteristics. As in the pitch realm, the transposition operations involve adding a constant to the value of a sound color on a given dimension. The constant may be positive or negative, so the transposition may be "up" or "down." We must include a specification of the particular dimension whose value is to be changed, and we must remember that the acoustic correlates of the operation, in terms of changes in F_1 and F_2, are dictated by the psychoacoustic relationship between changes in those two resonance frequencies and change in the (perceptual) value of the sound color with respect to the dimension in question. It appears, therefore, that transposition in sound color and pitch are close enough in principle to justify use of similar names for the two operations.

Figure 16 illustrates the transposition of a sequence of three colors, [uu oe oo], with respect to each of the three dimensions of ACUTENESS, OPENNESS, and SMALLNESS. The sequence of colors is translated in each case in a direction perpendicular to the equal-value contours of the dimension in question. Although each transposed sequence consists of colors quite different from the original sequence, the size and direction of the distances between the colors with respect to the particular transposed dimension are preserved.

RECORDED EXAMPLES 3A, 3B, 3C

Focusing, for illustrative purposes, on the ACUTENESS transposition, let us examine in detail the distance relationships between the original and the transposed sequence. Between [uu] and [oe], the first and second members of the original sequence, is a positive change from a large negative ACUTENESS value to zero ACUTENESS. This is followed by a change back to the negative ACUTENESS value of [oo], the third member of the original sequence. The transposed sequence, approximately [oe ii ne], involves changes in ACUTENESS of the same directions and distances, but now varying between zero and positive values of ACUTENESS.

The ACUTENESS transposition preserves OPENNESS values of the original sequence: the negative OPENNESS values of the first two members of the sequence are the same in the transposition, as is the zero OPENNESS of the last color. This is a consequence of the orthogonality of ACUTENESS and OPENNESS, however, and not a universal property of color transposition. We can see, in this case, that the relationships

A THEORY OF SOUND COLOR

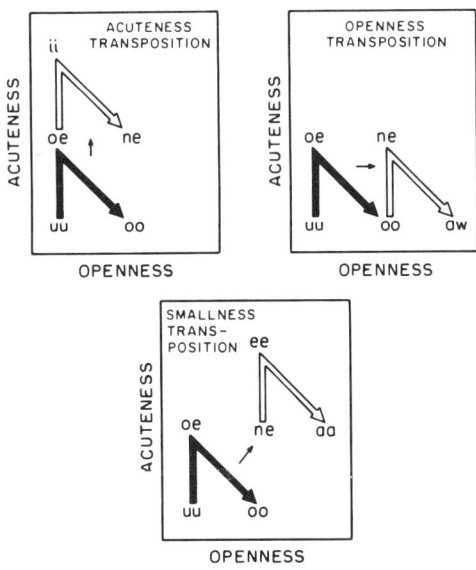

FIGURE 16: *Color transposition in the dimensions of ACUTENESS, OPENNESS, and SMALLNESS.*

among the SMALLNESS and LAXNESS values of the original sequence are not preserved.

LAXNESS Transposition

Transposition in LAXNESS is a somewhat different matter. Let us apply a LAXNESS transposition to to the sequence [uu oe ih (as in *bit*)]. As illustrated in Figure 17, a positive transposition in LAXNESS moves the original sequence toward the center of the sound-color space. Between the colors [uu] and [oe] there is no change in LAXNESS—they both have large negative values of LAXNESS. Between [oe] and [ih], on the other hand, there is a moderate increase in LAXNESS. The distances between the members of the transposed sequence appear to be smaller than those in the original, but with respect to LAXNESS alone the relationships are preserved. In the transposed sequence, roughly [uh oeh (a short umlauted o) ne], this same pattern repeats: no difference in LAXNESS between the first two colors, a moderate increase in LAXNESS between the second and third. The relationship with respect to the

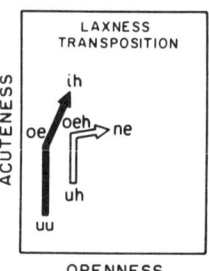

FIGURE 17: *LAXNESS transposition*.

other dimensions are drastically altered by the operation, however. For example, the [uu] to [uh] transformation represents an increase in each of the dimensions of ACUTENESS, OPENNESS, and SMALLNESS, whereas the [ih] to [ne] change represents a decrease in ACUTENESS, an increase in OPENNESS, and no change in SMALLNESS.

Wrap-Around in Color Transposition

The transpositions of the sound-color sequence illustrated in Figures 16 and 17 seem intuitively to be very similar to transposition of a series of pitches. Certain other sequences subjected to those same transpositions and certain other transpositions of the same sequence of colors present problems, however. For example, suppose we wished to transpose the sequence [uu oe oo] by a moderate amount in OPENNESS, but this time in a *negative* direction. What would we do with [uu] and [oe], whose OPENNESS is already at the most negative value possible in the space? Unless we are willing to limit transposition arbitrarily to those operations and sequences that do not move beyond the space, we must redefine the operation so that all colors can be transposed with respect to all the dimensions and by any amount. This redefinition is effected by what we may call *wrap-around*, a convention that has a rough analogy in pitch transposition.

The transposition of pitch classes as defined in atonal theory is carried out by addition *modulo 12*. Transposing a B♮ (pitch class 11) up a major third (4 semitones) carries one outside the pitch-class space to an E♭ in the second octave (pitch class 15 [11 + 4]), unless one subtracts the octave (12 semitones). In other words, 11 plus 4 must equal 3. In effect, the octave is "wrapped around" so that pitch class 0 follows pitch class 11.

In sound color there is no octave, but there is a similar barrier—the edge of the space. If, for each dimension, we imagine that one edge of the space is wrapped around to join the opposite edge, then any color that is transformed off one edge of the space appears, moving in the same direction, at the opposite edge.[34] A negative OPENNESS transposition of [uu], with wrap-around, converts that color into the highly OPEN [aw]. The same transposition of [oe] produces [aa]. A negative ACUTENESS transposition of [uu], however, produces the highly ACUTE [ii], and a negative SMALLNESS transposition of [uu] results in [ae]. If we start once more with our original sequence [uu oe oo] and OPENNESS-transpose it repeatedly in a positive direction, we must invoke the wrap-around provision as well. If the size of the transposition interval is a moderate one, we might produce in order the sequences [oo ne aw], [aw aa uu], and finally [uu oe oo] once more. Generalized transposition with respect to the dimensions of ACUTENESS and OPENNESS can be handled quite naturally using wrap-around. The distance relations among sets of colors—measured *modulo the space*—are preserved whether or not wrap-around must be invoked.

Wrap-Around in SMALLNESS Transposition

SMALLNESS wrap-around presents difficulties. The diamond shape of the SMALLNESS space causes generalized transposition—defined as addition, modulo "the edge of the space"—to distort the SMALLNESS distances between colors that wrap around. Suppose we repeatedly transposed our original sequence [uu oe oo] in a positive direction with respect to SMALLNESS. We would produce first [ne ee aa], an invariant transformation, but the second step would produce [ae oe oo], in which the SMALLNESS relationships are lost. It is not hard to verify that

---

[34] A striking analogy to wrap-around in sound-color transposition is found in certain video games in which an object encountering an edge of the screen suddenly appears on the opposite side of the screen.

the original sequence can be recovered by repeated SMALLNESS transpositions, but the operation no longer guarantees an invariant relation among the intermediate transposed colors.

One solution to this dilemma involves redefinition of the SMALLNESS dimension with equal-value contours that follow the diamond shape of the SMALLNESS space. Transposition in this new SMALLNESS space could still be defined as the addition of a constant, but the value of the constant would now be expressed in terms of a fixed number of equal-value contours over which the transposition was to take place. Figure 18 illustrates this alternative definition of the SMALLNESS dimension and transposition with respect to it.

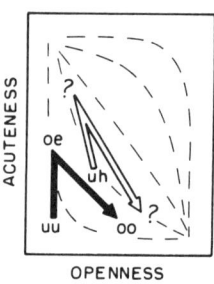

FIGURE 18: *Transposition with a redefined SMALLNESS dimension. The dotted lines represent new equal-value contours.*

The most serious problem with the redefined SMALLNESS dimension is the generation of new colors between [oe ee] and [oo aa] with simple transpositions that, for example, change [uu] to [ne]. The group structure of the colors in the space is upset. The significance of this consideration will be explored further in Chapter Seven. Here we must simply acknowledge that the transformation of musical elements into each other with closure is a fundamental and powerful feature of musical systems that should not be given up lightly.

In spite of its shortcomings, the first definition of SMALLNESS, with its parallel equal-value contours, permits transposition without introducing extraneous colors, and therefore it seems the best choice. Transposition in SMALLNESS must simply be acknowledged to be problematic. In the following discussion, then, SMALLNESS is the dimension as originally defined in Rule 2 and Figure 13. (The alternative, redefined SMALLNESS dimension may be of some use in musical contexts or styles that require continuous transposition and in which the group structure is not essential.)

LAXNESS Transposition with Wrap-Around

The group structure of a set of colors is not necessarily upset by LAXNESS transposition with wrap-around. However, as colors are transposed into the maximally (zero) LAX position, they lose their identity. Wrap-around in this case may be thought of, more intuitively, as wrap-"inside-out." In fact, the problem can be solved by tacitly tagging colors as they are transposed into the neutral position. Thus successive transpositions in the direction of increasing LAXNESS of the sequence [uu ii ae] would generate first a short form of the sequence, say [uh ih eh], then [ne(uh) ne(ih) ne(eh)], where the colors in parentheses tag the identical, maximally LAX, neutral colors with the directions from which they have been transposed, and finally—by virtue of wrap-inside-out—the original [uu ii ae]. It appears unnecessary to develop an elaborate theoretical structure or notation to handle this peculiarity of LAXNESS transposition; one need only keep track of the order positions of colors to make the proper transformations as a set emerges from the neutral point.

RECORDED EXAMPLE 4

The Transposition Rule: Rule 3a

The formal statement of the first operation on sound color follows directly from the foregoing discussion:

> *Rule 3a:* To transpose a sound color with respect to one dimension, add a constant to the value of the sound color on that dimension, modulo the edge of the space.

We can see that transposition depends intimately on the equal-value contours of the various dimensions for the specification of both the direction of change in color—i.e., the orthogonal direction—and its amount—i.e., what is meant by a "constant value." As I have indicated above, although psychoacoustic specification of the dimensions is not required, their relationships to each other must be specified for Rule 3a to be well defined.

Finally, it must be acknowledged that, when wrap-around must be invoked, sound-color transposition is not perceptually compelling. This is not to say that listeners cannot learn to follow such transpositions, but the task will undoubtedly prove difficult. On the other hand, transposition *without* wrap-around seems quite natural and easy to hear. An extension of that perceptually clear operation, wrap-around is necessary in order to preserve the group structure of sound-color transposition and is therefore justifiable on theoretical, if not perceptual, grounds.

Sound-Color Inversions: Rule 3b

Before the development of the twelve-tone system of pitch organization, inversion was a rare and rather specialized operation.[35] Its organizing force was apparently held to be inferior to that of transposition. However, as one of the fundamental operations in the twelve-tone system, *inversion* defined exactly is held, in that system and in certain other "atonal" musical systems, to be as legitimate as transposition. Nevertheless, there has been some criticism of the equal status of inversion (e.g., Lewin 1977) and it seems fair to suggest, on perceptual or intuitive grounds, that transposition is the stronger pitch operation.

The opposite is true in the realm of sound color. Here the operations of inversion, defined with respect to a particular dimension, generate none of the problems associated with transposition. Although undefined with respect to LAXNESS, inversion in the three remaining dimensions is simple in concept and perceptually compelling.

Let us begin by stating the second rule of transformation of sound color:

> *Rule 3b:* To invert a sound color with respect to one dimension, negate its value on that dimension.

[35] The term *inversion* is used here not as an indicator of the member of a triad that is in the bass, as in "first inversion chord," but in the sense of "inverting a melody" or "inverting a pitch-class set."

Inversion, like transposition, requires that the values of sound color on each dimension be specified, but the acoustic correlates need not be. In other words, both transposition and inversion are operations entirely in perceptual space. The key to the inversion operation is the location of the zero equal-value contour on each dimension and, therefore, the location of the maximally LAX position. Color can be said to be inverted around axes defined by the neutral color.

Let us take as an example the sequence [oo cc uu aw] and ask what the inversions would be with respect to the dimensions of ACUTENESS, OPENNESS, and SMALLNESS. Figure 19 illustrates these operations.

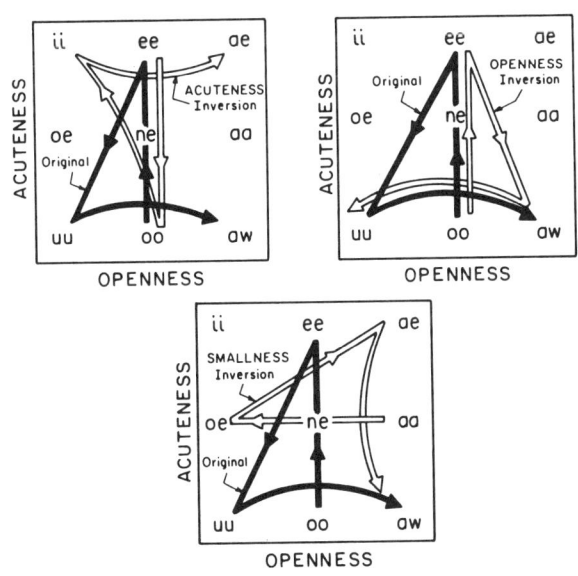

FIGURE 19: *Color inversion in the dimensions of ACUTENESS, OPENNESS, and SMALLNESS.*

ACUTENESS Inversion

The *gravity*, or negative ACUTENESS, of most of the sequence in Figure 19 is converted by the ACUTENESS inversion into the sequence [ee oo ii ae], whose overall character is strongly positive in ACUTENESS. The size of the ACUTENESS intervals from one member of the sequence to the next is preserved by the inversion, whereas the directions of the intervals are reversed. As we might expect from the orthogonality of

RECORDED EXAMPLE 5A

ACUTENESS and OPENNESS, the OPENNESS relations among the members of the sequence remain the same in the ACUTENESS inversion. Again not unexpectedly, the SMALLNESS relations among the colors in the sequence are disturbed in the transformed sequence.

OPENNESS Inversion

RECORDED EXAMPLE 5B

The effect of the OPENNESS inversion is analogous to that of the ACUTENESS inversion, but because of the symmetry of the sequence about the zero OPENNESS contour the inversion, [oo ee aw uu], is made up of the same colors as the original but in a different order. Since both [oo] and [ee] fall on the zero OPENNESS contour, they invert into themselves. The exchange between [uu] and [aw] preserves the size of the OPENNESS intervals between the second and third and the third and fourth colors and reverses their direction. As we would expect, the ACUTENESS relations among the members of the sequence are unchanged and the SMALLNESS relations are modified.

SMALLNESS Inversion

RECORDED EXAMPLE 5C

Inversion with respect to SMALLNESS raises none of the difficulties of transposition in that dimension. The SMALLNESS inversion of [ee oo ii ae], the sequence [aa oe ae aw], is a sort of rotation of the original that preserves SMALLNESS distances and inverts directions. The ACUTENESS and OPENNESS relationships are upset, as we would expect.

Effects of Inversions on LAXNESS

Although inversion is not defined with regard to LAXNESS, we can ask what effect inversions on the other dimensions have on LAXNESS. Since the original sequence had only colors of strongly negative LAXNESS, it is unrevealing with respect to that question; all the inverted sequences are made up of colors of LAXNESS as negative as the original. Since, however, the inversion operations by definition involve reflections about the neutral point, we can see that short colors—those medium in LAXNESS—will always invert into other short colors. We can conclude, then, that all inversions leave LAXNESS invariant.

Operations in "Shifted" Spaces

When the color space is contracted or expanded in response to a change in the overall length or size of the filter, the neutral point shifts along with the other colors in the space. Even though the neutral point has been moved away from the "normal" position ($F_1 = 500$ Hz and F_2 1500 Hz) and the F_1–F_2 loci of the axes about which the inversions are defined have been changed, the definitions of the inversion operations need not be changed, nor need we alter the transposition operations. This is because the shifts affect only the *acoustic* correlates of the sound-color space and do not alter the relationships among the dimensions in the "normalized" perceptual space of sound color.

To clarify the situation, let us be a bit more specific here about the process of normalization. Let us suppose that when first hearing a sequence of colors, a listener's auditory system calculates—largely on the basis of the frequencies of the higher resonances—the overall size of the filter producing the sound. Let us suppose further that the result of this calculation is expressed as a location in the F_1–F_2 plane of the point of maximal LAXNESS and the equal-value contours of all the dimensions. If the neutral point is normal—that is, at F_1 equal to 500 Hz and F_2 equal to 1500 Hz—no adjustments are required; the operations can be carried out on the basis of Rules 3a and 3b. If, on the other hand, the filter is larger or smaller than normal and the location of the neutral point is judged to be at values of F_1 and F_2 that are lower or higher than 500 and 1500 Hz, then the normalization process is invoked. A correction factor must be applied to the perceived sound to adjust it back to normal. Now the inversion operations will act on the adjusted sound colors exactly as they do on the normal colors. It may be that the actual psychological process involves changes in the operations to fit the filter rather than adjustments of the incoming stimulus. In either case, Rules 3a and 3b can stand as stated. Their effects will be the same whether a sound-color space is normal or shifted.

The Rationale: Are Color Transposition and Inversion Real?

Are the transposition and inversion operations on sound color legitimate, perceivable, natural? One way of answering this question is to suggest that color transposition and inversion are as legitimate, perceivable, and natural as the analogous pitch operations.

If we mean by "legitimate" that formal or "lawful" properties of sets must be preserved, then transposition with respect to ACUTENESS, OPENNESS, and LAXNESS and inversion with respect to ACUTENESS, OPENNESS and SMALLNESS satisfy that condition. In all color operations except the troublesome SMALLNESS transposition, the sound-color intervals are preserved.

If we mean by "perceivable" that the color invariances must be directly and immediately apparent to a listener of little sophistication, we must answer more equivocally. Sometimes both the transposition and the inversion of color sequences are quite direct and clear—the ACUTENESS inversion of the sequence [oo ee uu aw], cited above, appears to be such a case. Other color operations—for example, the SMALLNESS inversion of that same sequence and any transposition that requires wrap-around—seem less direct. This same equivocation must be maintained in the evaluation of the perceptibility of pitch transpositions and inversions. Few would deny that some transposed or inverted pitch-sets seem more closely related to their prime forms than others. To detect the invariance in all pitch operations, either some thought about the heard sound sequence is required of the listener, or a special mode of listening that requires training must be maintained. It is certainly not farfetched to maintain that processes of no greater complexity are required to detect invariances in color sets subjected to transposition and inversion than are required for detecting invariances in transposed or inverted pitch classes.

If we mean by "natural" a *predilection* to transpose or invert colors, we must again be equivocal, but hardly more equivocal than in evaluating the predilections for the analogous pitch operations. The general form of transposition without wrap-around—a simple shift along a perceptual dimension—is so nearly ubiquitous in human sensory processes that we can consider that operation to be quite natural without further argument. On the other hand, the breaks in the "shape" of a sequence of colors when wrap-around must be invoked are analogous to the changes in profile that follow from modulo-12 transposition of pitch classes. Here the realization of a pitch-class transposition in specific pitches, in which the composer can choose whether to preserve the pitch profile, presents an advantage over sound-color transposition, but at the level of pitch classes and colors the two realms are similar.

Inversion is a harder case, but it is equally difficult in pitch and color. Psychologists have identified a tendency for human beings to equate mirror images and to perform mental operations like those required for color and pitch inversion. Evidence for these kinds of operations will be discussed in Chapter Five. Our common experience with inverted melodies and various kinds of symmetrical visual figures hints, at least, that inversion along many sensory dimensions may be a biologically natural operation.

The Artificiality of the Sound-Color Operations

Even if we had no biological predilection, we should not be overly concerned about introducing "unnatural" operations into music. In the history of European concert music, essentially all the developments in musical structure have been artificial. They have grown out of previous developments by a certain logic, but those logical developments were not inevitable. The varieties of musical phenomena in the cultures of the world argue strongly in favor of artifice as the driving force. Just as the notion of inverting a melody and, later, a pitch-class set grew artificially out of previous musical practice, the inversion of sound color can be said to grow artificially out of previous practice—the inversion of pitch. Even if we hold sound-color operations to be less than natural (in the sense, that is, of following from biological necessity) we must come to the same conclusion about most of the well-established operations upon which musical structure depends.

Why Is the Neutral Color the Axis of Inversion?

If we accept sound-color inversion as a possible, perhaps desirable, operation, we still must ask whether the particular kinds of inversion specified in Rule 3b are the correct or most logical choices. The central issue is whether the maximally LAX point is the proper axis of inversion.

It seems reasonable to ask that sound-color inversion operations derived from existing pitch theory should share certain general characteristics with pitch inversion. We have already mentioned, in connection with the discussion of wrap-around in SMALLNESS transposition, the desirability of transforming a domain onto itself. The inversion of a twelve-tone pitch set is simply a reordering of the same twelve pitch classes; we would like to maintain a similar relationship between certain "complete" color sets and their inversions.[36] If we retain the wrap-around feature at the edges of the space, this requirement constrains the position of the axis very little. If we eschew that construct, then only an axis in the center of the space will do.

This can be shown by considering the ACUTENESS inversion of the original sequence, [oo ee uu aw], about some other axis. If the axis for ACUTENESS inversion were an equal-ACUTENESS contour that fell, say, between the [oe aa] and the [ii ae] contours, the second color of the sequence, [ee], would transform into [ne]. The disposition of the remaining three colors would depend on whether a kind of wrap-around was invoked. Without it, [oo], [uu], and [aw] would be inverted outside the space into hyper-ACUTENESS locations above the upper edge of the space. The group structure of the inversion operations would be lost.

With wrap-around, the three remaining colors would invert into themselves (see Figure 20). With this off-center axis and an assumption of wrap-around, the ACUTENESS inversion of [oo ee uu aw] would be [oo ne uu aw]. Now, however, another general requirement on the inversion operation (hinted at above) can be seen to be violated. The ACUTENESS intervals between the set members are not preserved by the higher-than-normal axis of inversion with wrap-around. It is easy to verify that all such off-center axes, when invoked with wrap-around, cause inversion in all the dimensions to distort the distance relations among the colors.

This leads us to ask what axes, aside from the neutral contour, we could use that would preserve both conditions. The very edges of the

---

[36]The criteria for these kinds of sound-color sets are discussed in Chapter Seven.

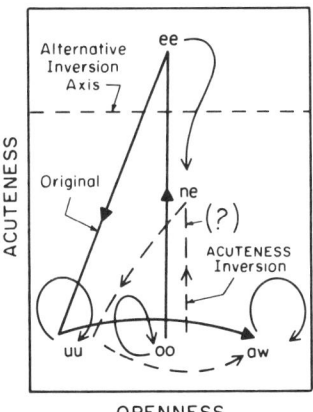

FIGURE 20: *ACUTENESS inversion with an axis above the neutral contour. If we assume that [ee] in the original sequence is 0.5 units above the axis, then the inversion is 0.5 units below the axis. The remaining members of the original sequence can be said to begin 1.5 units below the proposed axis. The negation of the ACUTENESS value is accounted for with 0.5 units to the [ii ae] contour, 0.5 units to the top edge of the space, and 0.5 units up from the bottom of the space to the [uu aw] contour.*

sound-color space suggest themselves. In such a case, inversion with respect to a dimension might be construed as reflection from, say, the edge having maximum values. It is easy to confirm that this process would always yield the original sequence; inversion, defined in this way, would be no operation at all. Alternatively, inversion around the edge of the space might be defined *with wrap-around*. Then low values on a dimension would invert to high values. For example, ACUTENESS inversion, defined about either the topmost or the bottommost equal-ACUTENESS contour, would transform [oo ee uu aw] into [ee oo ii ae], preserving the ACUTENESS distances. But this is exactly the same sequence as that produced by the original definition of ACUTENESS inversion, with its axis on the neutral equal-ACUTENESS contour. Apparently, negation with respect to the edges of the space is either an identity operation or it is equivalent to negation about the center. We are led back, therefore, to the original definitions of the inversions, both because they do not require appeal to the wrap-around feature and because all the inversion axes are easily specified in terms of the single neutral point.

If we are to have operations of inversion in the sound-color realm, the most straightforward approach seems to be the one specified in Rule 3b. The dimensional structure of the space is acknowledged by that rule, as is the analogy with inversion in pitch. There appear to be grounds for assuming that we are capable of learning to perceive the relationships between the inversions and their prime forms. The desirable properties of "onto" mapping and preservation of distance relations apply to all the inversion operations.

Combinations of Operations

Clearly, composers are concerned with the effects of performing successive operations on sound-color sets. The regularities of such combinations of sound-color operations, however, can be regarded most properly as a matter of compositional application of the theory rather than a part of the theory proper; as such, they will be dealt with in Chapter Seven.[37]

Extensions of the Operations

A few remaining issues present theoretical problems that are both psychological and musical in nature. These topics are beyond the scope of the theory itself, but they arise naturally from it.

---

[37] In the form stated in Rules 1, 2, 3a, and 3b, the sound-color theory is susceptible to appropriate scientific verification or falsification. Rules 3a and 3b are closer to music-theoretical claims than to psychological theories, but even their assertions of invariance may be interpreted psychologically and investigated with psychological methods. Musical structure too may be investigatible with scientific methods, but the chances are that well-known music, or at least well-known styles of music, would provide the best subject for productive research in that field (e.g., Deutsch 1978). The new musical structures that may grow out of the theory of sound color are unlikely to be of much scientific interest. The distinction between scientifically verifiable claims of the theory and its musical application is admittedly somewhat arbitrary, but in an attempt to emphasize the dual character of the theory, this chapter is concerned with the presentation of aspects of the theory that can be construed as psychological as well as musical. Developments of the theory that seem more exclusively of interest to composers will be saved for Chapter Seven.

Color Hierarchies

Heinrich Schenker has shown how layers of organization grow out of the recursive application of the fundamental rules of voice-leading.[38] No hierarchical system of comparable force follows from twelve-tone or atonal theory. Composers of our time work out large-scale pitch structures in any of a wide variety of ways. The structures that result may be no less convincing than those of tonal music, but in atonal and twelve-tone music the organization of pitch structure on the large scale is more a product of the composer's individual imagination and less a necessity of the system itself.

The theory of sound color, like contemporary pitch theories, does not prescribe a large-scale structure. Rather, it provides a context in which alternative hierarchies can be constructed. In a somewhat stronger sense than pitch, however, the fundamental dependence of sound color on a weakly coupled source/filter system suggests a hierarchy inherent in the sound itself. Let us examine certain of these suggestions.

Hierarchies by Means of Successive Filtering

If we color a sound by filtering it, we can presumably treat the result of that process as a source for a second filter system, and the output of that system as the input for a third filter, and so on. Remembering that one effect of resonances in the filter is a negative slope to the overall spectrum envelope with frequency, we must recognize that there is a practical limit to this hierarchical process. The progressive loss of high frequencies can, however, be compensated artificially by suitable introduction of higher pole corrections. By exploiting such successive filtering a potentially powerful means of building musical structure can be explored.

The differentiation of the levels of sound color in successive filtering can be reinforced by maintaining differences in timing among the

---

[38] Many of the writings of Schenker, including his *Free Composition* (1979), are now available in English. Two annotated bibliographies by David Beach (1969, 1979) are useful guides to Schenker's own publications and writings by others who are inspired by, and in some cases critical of, his theories.

levels. Let us suppose, for example, that a color pattern following the original sequence, [oo ee uu aw], is imposed on a source that is strong in high frequencies. Let us suppose further that that sound is then passed through a second filter whose resonance peaks are fairly broad and constant at frequencies close to those of the [oo] color. The retrograde of the original sequence might then follow, with the second filter shifting to an [ee]-like color. A third transformation of the original sequence, say the ACUTENESS inversion of the retrograde, would follow, with the second filter moving to [uu]. Finally, following that sequence would be the ACUTENESS inversion of the original form of the sequence with the second filter set at an [aw] color. The slower-moving second filter in this example would express the original sequence once while, concurrently, four versions of the sequence are presented in the first filter.

One result of this kind of double filtering is reinforcement of the amplitude of colors that occur simultaneously in both filters, an emphasis that can serve as a clue to the sound-color structure. Alternatively, a transformation of the faster sequence—say, SMALLNESS inversion—can be chosen that does not duplicate the spectrum envelope of the second filter.

The use of simultaneous filter hierarchies will undoubtedly raise a number of compositional issues that are beyond the scope of the present study. The fact remains that multiple filtering, organized according to sound-color dimensions and operations, is a natural outgrowth of the basic theory.

Hierarchies by Sound-Color Space Limitations

Shifting of the sound-color space, an alteration that invokes the process of normalization, can be adapted to produce another kind of hierarchical structure. Shifts were introduced as a means of contracting or expanding the sound-color space in a specific way to account for tubes of overall lengths that are greater than or less than the standard 17 centimeters of the adult male vocal tract. We can generalize the concept of shifting to encompass a set of characteristic limitations of the sound-color space that can themselves express a sound-color profile.

Figure 21 illustrates a succession of subspaces that follows the original sequence, [oo ee uu aw]. Within each of these subspaces is presented a "space-limited" version of the same sequence. As in multiple filtering, this kind of organization probably requires temporal contrasts to distinguish the levels of the hierarchy. It does not appear to be as sensitive to the actual filter characteristics as does successive filtering. The degree to which the space can be limited without upsetting the perceived invariances in the sound-color sequences must be determined by further study.

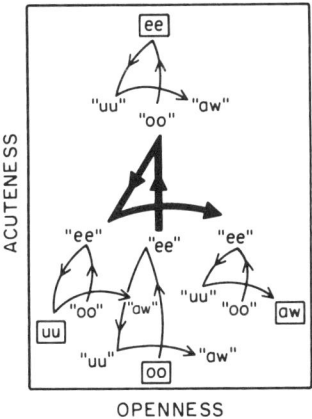

FIGURE 21: *Hierarchies by means of subspaces. The sequence [oo ee uu aw] appears in subspaces surrounding the colors shown in boxes.*

Color Dynamics

One controversial issue in the distinctive-feature theory is the appropriateness of calling certain consonantal features by the same names as those of vowels (Fant 1973b). An extension of the theory of sound color into the realm of moving spectrum envelopes also must be considered controversial. Nevertheless, we can imagine that musically relevant invariances may arise in consonant-like sounds that derive from steady-state sound colors. The class of consonants known as the liquids (the initial sound in "will" or in "you," for example) are very commonly treated by acoustic phoneticians as falling on a continuum from vowels to diphthongs to liquids to stops (Liberman et al. 1956). The critical parameter in this continuum is the speed with which the F-pattern changes.

Leaving for the future an investigation of the temporal limitations on sound-color identification, we can well imagine the expression of a sequence of colors by means of "liquids" whose F-pattern begins at one color and moves quickly to another.[39] The transition in the word "you," for example, is from the color [ii] to the color [uu]. The sound-color operations of transposition and inversion could be employed on these kinds of sounds as well.

Clearly, the theory of sound color can be applied to the control of more slowly moving F-patterns. In the same sense that the end points of glissandi are counted in serial music as pitch classes, we can treat moving sound colors as legitimate expressions of a sound-color structure. There is nothing to prevent one from, for example, inverting with respect to OPENNESS a continuously changing sound-color pattern that moves from [oo] to [ee] to [uu] to [aw]. The result would be a "color glissando," [oo ee aw uu].

Color Mixture

The theory of sound color does not deal explicitly with sound-color mixture.[40] No rules analogous to those defining dissonance and consonance in tonal music are presented that limit the simultaneous combination of sound colors. A few observations on the subject can be made, however, that follow in some sense from the theory.

Questions about the relationships between the sources in sounds having mixed colors seem to be critical. If the spectrum envelopes of two colors are combined and excited by the same source, it is likely, as a general rule, that the sound color of the combination will be quite different from the colors of the two components. On the other hand, if the sources for two different filters are drastically different in charac-

---

[39]Empirical evidence from speech research indicates a close relationship between F-pattern movements and movements in the speech cavity (e.g., Ladefoged and Harshman 1979).

[40]Yilmaz (1967) has proposed an interesting theory of vowel perception that predicts effects similar to those of sound-color mixture.

ter, it is likely that the sound colors will not blend but, rather, the two filters will be heard as separate "voices." We can suppose that this tendency to separate simultaneous sounds having different colors into contrapuntal lines would be enhanced by temporal contrasts in the source events and would be reduced by temporal identity. To confirm or deny these observations, empirical study of the simultaneous combination of sound colors will be required. The attempt by Cogan and Escot (1976) to treat added colors as simply equal, in some sense, to the sum of the parts is no doubt an oversimplification.

Possible Additional Dimensions

The dimensions of sound color as presented in this chapter are not necessarily a complete set.[41] Among the possible additional dimensions that may be found useful are HARDNESS, a correlate of the narrowness of the resonance peaks, and BRIGHTNESS, a dimension related to the frequency of the third resonance. However, there seems to be sufficient reason for regarding the four dimensions identified in the theory proper as the primary attributes of sound color. They provide a field rich enough in possibilities to make some headway in the attempt to bring to the control of sound color the kind of "lawfulness" to which this study is devoted.

Summary of the Sound-Color Rules

The rules that make up the theory of sound color are restated here as a convenient reference and a summary of Chapter Two.

Rule 1: To keep the color of a sound constant, keep the resonance frequencies (F_1, F_2, F_3, etc.) and bandwidths (B_1, B_2, B_3, etc.) of the filter constant.

Rule 2: Sound color has the dimensions of OPENNESS, ACUTENESS, LAXNESS, and SMALLNESS. These dimensions

---

[41] Within themselves, however, they form a unified whole, as will be shown in Chapter Seven.

are defined by the equal-value contours shown in Figure 13. To hold sound color constant with respect to one dimension, change the values of F_1 and F_2 in such a way as to remain on one of that dimension's equal-value contours.

Rule 3a: To transpose a sound color with respect to one dimension, add a constant to the value of the sound color on that dimension, modulo the edge of the space.

Rule 3b: To invert a sound color with respect to one dimension, negate its value on that dimension.

These are the main assertions of the theory. It has been presented in this chapter largely from a rational point of view. In the following chapters, empirical evidence will be reviewed from a number of fields in an attempt to give further support to the theory.

CHAPTER THREE

Evidence from Auditory Physiology

PHYSIOLOGISTS BRING TO their study of the auditory system methods that have the ring of objectivity. Recordings from auditory neurons, direct measurements of the mechanical actions of the inner ear, or tracings of the connections between various centers of the auditory nervous system seem more concrete, more certain, than the methods of the other sciences concerned with hearing. This "hard" scientific character motivates us to turn to physiology in search of scientific support for the theory of sound color.

However, the very precision and concreteness that attract us in physiological research serve to limit its applicability to questions about sound color. Rule 1, for example, is stated in psychological terms; it is a hypothesis about the auditory experience of listeners that makes no claim about *where* in the auditory system sounds with similar F-patterns are classified as equivalent or *how* such a classification takes place. Yet auditory physiology is nearly always concerned with exactly where and how some particular action takes place. The questions physiologists ask are not the same as those asked by psychologists and, as a result, physiological research appears often to be somewhat tangential to the theory of sound color.

A more serious problem arises from the great complexity of the auditory system and the relatively poor state of our knowledge about it. Suppose we try to turn knowledge of physiology into direct support for, or negation of, psychological (or music) theory by asking, not where or how some function is performed in the auditory system, but whether it is performed at all. If, for example, physiologists could assure us that similar F-patterns excited by different sources are never treated as if they were equivalent anywhere in the auditory system, then we would have to give up Rule 1 and the theory of sound color would be disproved. Unfortunately, clear-cut negative assertions of this sort can hardly ever be derived from physiological results. With very few exceptions, each stage in the processing of a sound stimulus by the auditory system has been found to involve many structures of both similar and different morphologies and apparent functions. Knowledge of the workings of these structures, their organization within various organs and brain centers, and the connections among the parts of the system is relatively recent, more or less controversial, and certainly incomplete. A failure to find some function in a particular auditory center gives no guarantee that investigation of some other center, another portion of that same center, or even, under different experimental conditions, the specific structure under study will not demonstrate that function. Moreover, particular negative results do not rule out certain functions as resulting from the systematic interactions of the various auditory centers (cf. Haugeland 1981). There may come a time when mammalian auditory systems are understood well enough so that certain negative results can be interpreted as disconfirming a psychological claim. At the moment and for at least the near future, however, it is possible to disprove only the most implausible of psychological theories by means of physiological research.

Positive physiological results are a different matter. Finding a structure in the auditory system that performs the function predicted by a psychological assertion (the analysis of F-patterns, for example) tends to support that assertion. We cannot prove that the structure in question is the sole, or even the most important, mechanism underlying the asserted psychological process, but at least a candidate for that mechanism has been identified. Moreover, other facts about the physiological experiment, such as the location of the structure in the auditory system and the phylogenetic relationship of the experimental animal to man, can give us indications of how generally distributed—roughly speaking, how "primitive"—the process may be and thus can suggest the extent to which it is free from species-specific modes of listening.

In attempting to deal with these problems of interpretation we must grant from the outset that physiology and psychoacoustics (and, for that matter, cognitive psychology, music theory, composition, etc.) are different fields, each with its own questions and its own methods. Even though these fields study the same system and have the same ultimate goal of understanding what and how we hear, there are logical and practical impediments to answering questions in one of these fields with research in another. Nevertheless, the possibility of identifying positive physiological evidence for the theory of sound color justifies a review of relevant work in that field. If structures have been found in mammalian auditory systems that appear to analyze resonance patterns, to treat filter characteristics differently from source characteristics, or to differentiate the dimensions of sound color,[1] we can interpret such findings as strong suggestive support for the theory of sound color.

ANATOMY OF THE AUDITORY SYSTEM

The auditory system is made up of a mechanical section—most significantly, the inner ear—and a neural section. The neural section begins at the nerve endings in the inner ear and can be traced through several centers, or nuclei, from the more peripheral—those close to the inner ear—to the more central—those closer to the cerebral cortex.

The Cochlea

The input to the auditory system is most properly viewed, not as the waveform of a sound but, rather, as a frequency analysis of it that is

---

[1] Rules 3a and 3b deal with a level of complexity that is well beyond the bounds of our present physiological knowledge. Nothing that we know of auditory physiology denies that sound-color operations are possible, but only the flimsiest and most indirect of suggestions pertaining to them can be derived from physiological research.

carried out in the fluid-filled, snail-shaped *cochlea* of the inner ear. The cochlea is divided throughout its length by a partition containing the *organ of Corti*, in which are the terminal neurons of the auditory nerve. At the bottom of the cochlear partition supporting the organ of Corti is the *basilar membrane*, which varies regularly in width from narrow, and hence stiff, at the end of the cochlea nearest the middle ear (the base) to wide, and hence flexible, at the opposite end (the apex). As the elegant experiments of Georg von Békésy show, a sound wave is transformed in the cochlea into a displacement of the basilar membrane in the form of a traveling wave. The wave rises to a maximum amplitude at different places depending on frequency—high frequencies near the narrow basal end and low frequencies near the apical end (Békésy 1960).

Since the first-order neurons of the auditory nerve are distributed along the cochlea, it has long been presumed that those nearest the base encode high-frequency sounds and those nearest the apex encode low-frequency sounds. In other words, frequency in the sound is converted into position along the basilar membrane and across the population of auditory neurons. Just exactly how frequency is encoded in the auditory nerve has been a matter of intense study over many years and is controversial (Eldredge 1974), but most researchers agree that a combination of mechanical and neural processes at the periphery of the auditory system present to the higher levels of the auditory system, not the waveform of the sound, but some representation of its frequency spectrum. The term *tonotopic* refers to the association of frequency with position along the cochlea; that term and the more physiologically neutral *cochleotopic* are used by physiologists to describe this representation of sounds in terms of frequency in the auditory nervous system.

The Pathways of the Auditory Nervous System

The neural pathways of the auditory system—the routes by which the first neural representations of sounds are transmitted to successively higher centers of the brain—are complex, and our knowledge of them has been drastically revised only recently.[2] The first-order neurons twist together at the cochlea to form the auditory nerve, and they synapse for the first time in the *cochlear nucleus*. The majority of the neurons from the cochlear nucleus cross the midline to structures on the opposite side of the brain stem called the *olivary nuclei*. Above the olivary nuclei there are both *ipsilateral* (same-side) and *contralateral* (crossed) projections to the *inferior colliculae* in the midbrain. The next auditory center is the *medial geniculate* in the thalamus, which receives both contralateral and ipsilateral input. The final, highest level of the auditory system is the *auditory cortex*, an area within and around a crevice, called the Sylvian sulcus, in the temporal lobe of the brain.

Up to the level of the inferior colliculus, the auditory areas have long been known to exhibit cochleotopic organization (Whitfield 1967)—that is to say, there are topographical correspondences between location along the cochleus and location across neural centers. Until recently this organization has been regarded as less sure at the higher levels, particularly in the auditory cortex (Goldstein and Abeles 1975). Recent work, notably by Merzenich and his coworkers (Merzenich et al. 1977), has revealed representations of the cochlea throughout *all* levels of the auditory system (see Figure 22). Within each of the auditory nuclei, including the medial geniculate and the cortex, lines of nerve cells spatially distributed across the nuclei can be identified that are associated with specific positions along the cochlea. The new research has shown not only that the cochlea is represented spatially in the cortex but that it is represented there repeatedly in different areas and that these multiple representations correspond to multiple, cochleotopically organized nuclei in the lower centers as well (Merzenich et al. 1977). The auditory nervous system, like other sensory systems (Kaas et al. 1979), appears to be made up of a number of parallel tracks, each of which carries representations of the sensory surface. As is suggested by recent work of M. C. Liberman (1982), these tracks may be differentiated as peripherally as the first-order

---

[2] For the traditional view of these pathways, see Yost and Nielson (1977).

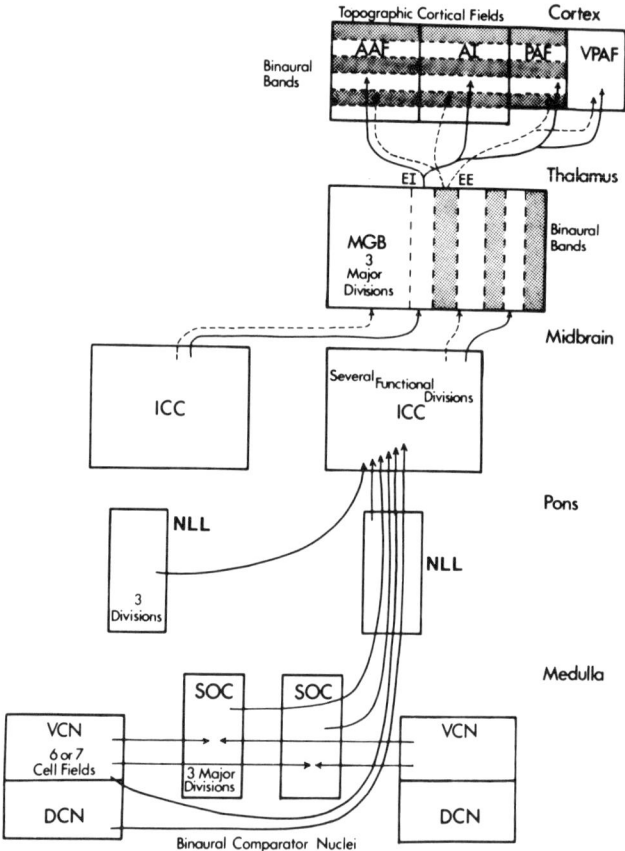

FIGURE 22: *A new conception of the auditory pathways. Drawn by Merzenich and Kaas (1980) from recent work indicating cochleotopic representations and both parallel and convergent projections among the levels of the auditory system. The neurons from the cochlea synapse in the cochlear nucleus. Neurons from the ventral cochlear nucleus (VCN) project to the superior olivary complex (SOC). Neurons from the ventral and dorsal cochlear nucleus (DCN) and from the olivary complex project to the nuclei of the lateral lemnisci (NLL) and, in turn, divisions from each of these sources project to the inferior colliculi (ICC). In separate, parallel tracks, neurons from the inferior colliculi connect to divisions of the medial geniculate bodies (MGB) in the thalamus. Finally, these divisions of the medial geniculate all project to at least four areas of the auditory cortex (AI, AAF, PAF, and VPAF).*

neurons in the cochlea, where differing populations of neurons have been identified with differing sensitivities.

The clearer, more complex picture of the layout of the auditory system painted by this recent research provides the background for a number of questions that grow out of the theory of sound color. Is one set of parallel connections devoted to analysis of filter characteristics and another to the source? Are there indications that the dimensions of sound color specified by Rule 2 of the theory may be differentiated somehow among the parallel tracks? Do any neural structures exhibit behavior suggesting that they may be involved in the normalization of F-patterns? Unfortunately, we cannot answer these questions. The anatomical findings are too recent, and in any case, auditory physiologists have not been concerned with questions like these. Some physiological research, however, has been done that pertains to the more basic parts of the theory.

FREQUENCY ANALYSIS IN THE PERIPHERY: CLASSICAL VIEWS

In the study of the functions of the auditory system, a central theme has been the manner in which the frequency analysis of sound is carried out in the more peripheral parts of the auditory system. Whitfield's *The Auditory Pathway* (1967) is a sophisticated, clear survey of the research in auditory physiology up to the mid 1960s. In a summary that interprets that research in terms of the sensory functions, he concentrates on how the peripheral parts of the auditory system must process sounds. Selecting a stimulus of particular interest—simultaneous tones of different frequencies—he asks what the neural discharge pattern in the cochlear nerve and the cochlear nucleus may look like in response to that sound. Figure 23a, taken from Whitfield's discussion, shows how, in response to the two different sinusoids, the pattern of excitation spreads across the array of neurons in the cochlea. The frequencies of the stimulus are represented as maxima in a broad pattern of activity. This representation of the two tones follows from consideration of the fairly coarse mechanical frequency analysis of the basilar

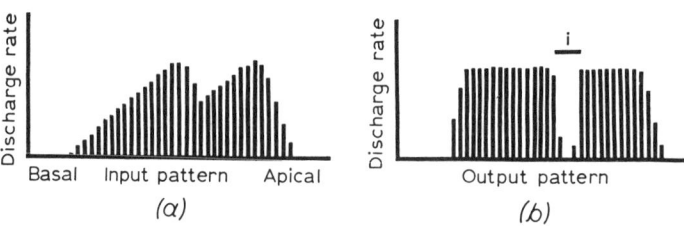

FIGURE 23: *Hypothetical response of the neurons in the cochlea (a) and the cochlear nucleus (b) to two simultaneous tones of different frequencies. (From Whitfield 1967.)*

membrane, the fact that the tones excite a fairly broad population of neurons, and the presence of neural sharpening.

Figure 23b gives Whitfield's conception of how the cochlear nucleus transforms the pattern reaching it from the cochlea. We see that the small dip in activity between the two maxima in the cochlear nerve has been deepened into a pronounced gap, whereas the maxima are broadened. These effects are brought about by saturation of the firing rates of neuron populations—spreading the peak—and by mutual suppression of the activity associated with one tone by the activity associated with the other—deepening the gap. The precise location of the peak of activity has been lost, but the distinction between the two tones—the fact that there were definitely two—is coded very effectively. Whitfield (1967, 156) suggests that this functioning of the peripheral auditory system forms the basis for distinguishing resonances in vowels. He argues that a similar spread of activity across the array of neurons and a similar sharpening of the gaps between the maxima in that pattern of activity would encode unmistakably the presence of resonances and—with some loss of precision—their frequencies.

The response of structures above the cochlear nucleus in the auditory portions of the central nervous system have been found to be increasingly complex and variable at successively higher levels (Whitfield 1967, 102). As of 1967 there was little unequivocal knowledge of the functions of those centers, and some of what was thought to be true then must now be qualified by the new anatomical findings. Nevertheless, one aim that can be detected throughout much of the research of that time has continued to inspire much work to the present day. This is the search for feature detectors.

In the visual-projection areas of the cortex, cells have been found that respond to particular shapes at particular locations in the visual field. Other cells respond to certain shapes moving in a certain direction at a given speed. The cells receive information from a broad region on the retina, encompassing the output of many neurons in the optic nerve (Hubel and Wiesel 1962). These, and cells whose favored stimuli are even more complex, can be said to be *feature detectors*: they seem to be "looking for" specific features within fairly broad areas of the visual field.

When these discoveries were made in the 1950s and early 1960s, auditory physiologists were motivated to seek analogous kinds of feature detectors in the higher centers of the auditory system. Although some cells were found that appeared to respond to characteristic patterns of sound (Kiang et al. 1965; Kiang 1968), the findings in the auditory system were much more ambiguous than in vision. The auditory cortex gave drastically different results depending on the state of arousal of the experimental animal, the location within the auditory cortex, and other experimental conditions (Brugge and Merzenich 1973).

These complications in the auditory cortex are explained in part by Merzenich et al. (1977), who cite marked differences between the organization of the auditory system, in which the sensory surface is represented in the brain centers by a line of cells, and the organization of other sensory systems, in which the sensory surface is represented by two-dimensional arrays of cells. But these unique characteristics of the auditory system do not rule out feature-detector cells. There have been successes in identifying and classifying feature detectors of a rather primitve kind in the cochlear nucleus (Rose et al. 1973; Møller 1977; Britt and Starr 1976) and, less surely, in the inferior colliculus and the medial geniculate bodies (Erulkar 1975). It seems unlikely that a characteristic of sensory systems as pervasive as feature detectors could be absent from the higher centers of the auditory system.

One question that always arises in the search for feature detectors is whether the favored stimuli of the particular cells under study have been found. Until very recently, physiologists have used as stimuli sounds that are simple mathematically. This choice was made largely in order to permit the physiological results to be generalized. Sinusoids were often used, for the same reason that they are used in investigating the response of acoustic or electronic filters. If we know how a system responds to sine waves of different frequencies, we can predict how it will respond to a complex sound that is made up of those frequencies. This assumption is valid, however, only in linear systems—those whose response strengths are proportional to the intensity of their inputs. And, as is suggested in Figure 23, the auditory system is highly non-linear. Recognizing this problem, auditory physiologists have recently begun to use natural sounds—for example, the vocalizations of the animal being studied (Manley and Mueller-Preuss 1981). Given this trend, we can hope for more success in finding species-specific feature detectors in the auditory system.

The concept of feature detection suggests sensory pre-programming for specific stimuli in an animal's environment. Studies that emphasize features, as we have seen, contrast strongly with studies using sine waves or other mathematically simple stimuli. A somewhat different approach can be taken that might be considered a middle course, retaining a measure of mathematical generalizability while recognizing the kinds of sounds that make a difference to the animal. If we analyze sounds in nature, including sounds made by animals, we find that many of them can be broken down into features that arise from a source and features that arise from a filter. The feature detectors in the visual system respond to somewhat abstract features—e.g., edges, corners, annuli—apparently not to full-fledged images of specific objects (Hubel and Wiesel 1962). It may be that filter resonances or entire F-patterns are among the auditory analogs to those abstract visual features. The use of animal models for understanding human auditory systems can be justified if, as was suggested in Chapter Two, we can regard filter characteristics as natural aspects of sound for the higher animals.

RESONANCE ANALYSIS

The hypothetical picture of analysis of frequency in the auditory periphery presented by Whitfield (1967) is compelling, but how accurate is it, and what evidence do we have that such analysis actually takes place? Whitfield himself cites some evidence, and more recent reviews point to considerably more (Evans 1974). Unfortunately, very little of that research deals directly with filtered sounds and resonances. To apply the model to the analysis of filters we would have to assume that the two-tone stimulus is something like that produced in a suitably excited two-resonance filter system.

Whitfield's model presents a problem that has been the object of considerable experimental research: the flattening and spreading of the discharge pattern of neurons as the intensity of the stimulus increases. The neural sharpening process distinguishes the two resonance peaks, but the saturation of the neurons blurs the resonance frequencies (see Figure 23b). Are the sharp peaks in the spectrum of the two tones recovered somehow in the neural responses? And do the responses to the two-tone stimuli resemble those to multiple-resonance systems? A series of recent studies gives us clear and convincing answers to these questions.

Vowels and the Auditory Nerve

Murray Sachs and Eric Young of Johns Hopkins Medical School have performed a series of elegant experiments on the responses of auditory nerve fibers to sounds made up of a buzz exciting a resonance system (Sachs and Young 1980; Young and Sachs 1979). Their stimuli were synthetic vowels with steady-state, periodic excitation and steady-state resonances. The resonance frequencies were typical of certain vowels but adjusted slightly so that the resonance peaks coincided with one of the harmonics of the source. Sachs and Young made use of a measure, not of the rate of firing of neurons at their characteristic frequencies[3]—the measure Whitfield (1967) and others have used—but of the average temporal intervals between firings (Young and Sachs 1979). The recordings from a broad population of neurons in the auditory nerve exhibited a pattern that strongly resembles the spectrum envelopes of the vowels. Prominent peaks at the frequencies of the first

[3]The characteristic frequency of an auditory neuron is the frequency to which it is most sensitive.

two resonances and less prominent peaks at the third resonance frequency can be seen clearly (see Figure 24). The results at high intensities of the vowel indicate that responses to the harmonics of the sound that are not near resonance frequencies are actually suppressed.

In a further study, the group at Johns Hopkins showed that the vowel spectrum envelope can be maintained in the temporal patterns of auditory nerve firings when the stimulus is masked by broad-band noise (Voigt, Sachs, and Young 1981). Those same researchers showed, in addition, that the spectrum envelope pattern is preserved in whispered vowels, i.e., when the source is noise (Voigt, Sachs, and Young 1982). This result is particularly strong support for the theory of sound color, because the source and the filter are in no way related in the case of noise excitation.[4]

The experiments of the Johns Hopkins group show that the auditory nerve itself, prior to any processing in the cochlear nucleus, exhibits a clear picture of the spectrum envelope, including sharp peaks at the frequencies of the formants in the stimulus. Thus, the acoustic correlate of the filter characteristics, the spectrum envelope, is shown to be preserved intact by processes in the cochlea. Rule 1″ of the theory, which is expressed in terms of spectrum envelope, is therefore strongly supported.

Other details of the results are relevant to the sound-color theory. First, we can observe that the peaks corresponding to the resonances are the invariant features of the auditory nerve representation: the off-peak components of the source are suppressed at high-intensity levels of the stimulus but not at low intensities. This invariance appears to provide physiological support for the argument—presented in Chapter Two in support of the final statement of Rule 1 in terms of F-pattern—that the peaks are more important perceptually than the valleys. Second, the third and higher resonances are less well defined in the auditory nerve representation than are the first two. This supports the assumption, also made in Chapter Two, that the effects of the higher

---

[4]Recent studies by the Johns Hopkins group indicate that dynamic shifts in resonance patterns, like those produced in certain consonants, are also represented in the auditory nerve (Miller and Sachs 1983).

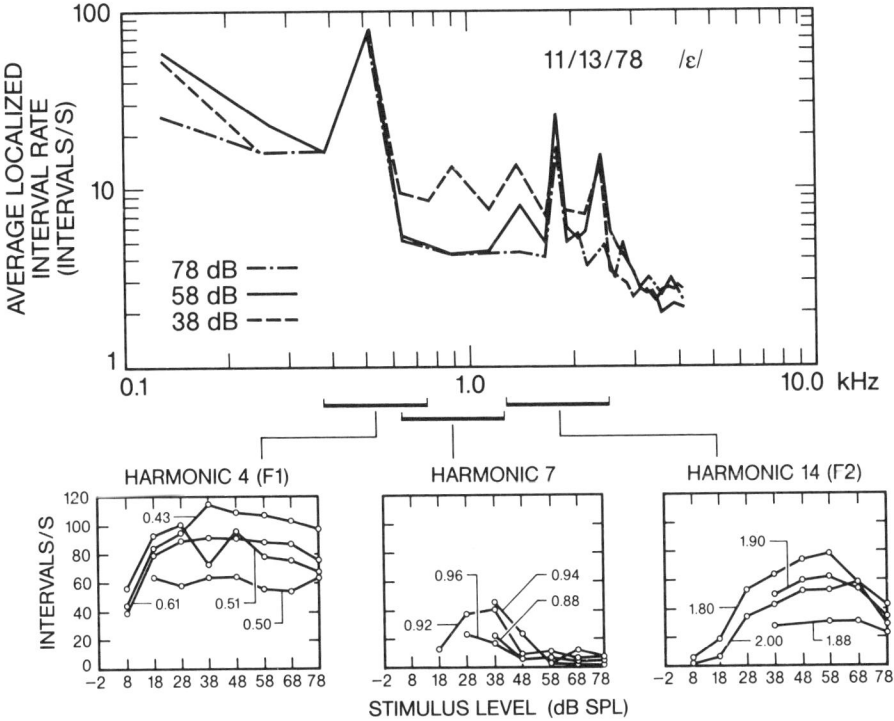

FIGURE 24: *Responses of neurons in the auditory nerve to the vowel [ɛ]. At all three intensities, the first two resonances have clear maxima. At high intensity (78 dB), the responses to the harmonics that fall between the first and second resonance are suppressed. (From Sachs and Young 1980.)*

resonances are secondary to those of F1 and F2. Finally, we can interpret the results as suggesting that the spectrum envelope—and possibly the frequencies of the first two resonances—may be analyzed peripherally. It follows, as was argued above, that the kind of invariance claimed in Rule 1 is a primitive auditory feature, not limited to the perception of speech. The stimuli used by the Johns Hopkins group may be regarded not simply as vowels but as examples of a broad class of filtered sounds. This more general, ethologically neutral attitude associates the results of Sachs, Young, and Voigt with precisely the domain of sounds covered by Rule 1.

The important studies of the Johns Hopkins group leave unanswered one question that pertains significantly to the sound-color theory. Rule 1 asserts the independence of the source and filter characteristics, so we are led to ask how neurons of the auditory nerve respond to sounds in which the excitation is in arbitrary relationship to the resonance frequencies. In particular, we would like to know whether there are indications of peaks in neural response occurring at resonance frequencies when the components of the source are off-peak or straddling the resonance frequency. If this strong test of the theory were to fail,[5] we would have to assume that some further analysis—perhaps some kind of template-matching process—is applied to the responses of the auditory neuron array, presumably at the level of the cochlear nucleus or above, to identify the resonance frequencies.

Resolution of Resonances

Studies of auditory nerve responses to noise with filter-like characteristics, although not concerned directly with the encoding of resonance patterns, throw additional light on the degree to which resonance peaks are resolved. Evans (1977) showed that comb-filtered noise—noise whose frequency spectrum exhibits peaks and valleys—is represented in the responses of nerve cells in the cochlear nucleus, although not in the cochlear nerve. As the peaks and valleys in the stimuli of that experiment were adjusted closer and closer, their representation in the cochlear nucleus began to deteriorate. This limit of resolution appears to corroborate the near-failure of differentiation between F_2 and F_3 in the results in Sachs and Young (1980) (see Figure 24).

Similar results were obtained by Smoorenberg and Linschoten (1977) with a stimulus consisting of a periodic complex tone made up of 63 equal-amplitude partials. The response of a neuron in the cochlear nucleus was studied with different frequencies of the stimulus. When the fundamental frequency of the complex tone was at the characteristic frequency of the neuron, the rate of response followed the fundamental waveform. As the complex tone was lowered in fre-

---

[5] In the case of the first formant, some results of Voigt, Sachs, and Young (1981) suggest that the response maxima may coincide with the source components rather than the resonance peaks.

quency, the neuron's firing pattern would follow successive harmonics of the stimulus up to the sixth harmonic. At this point the spacing of the harmonics became too narrow for the cell to differentiate individual components, and it began to respond in the form of a damped sinusoid at its own characteristic frequency. If we interpret the harmonics of the stimulus as if they were resonance peaks, this experiment too suggests a deterioration in the peak resolution comparable to that found by Evans (1977) and, by implication, Sachs and Young (1980).

Possible Cochlear-Nucleus Responses to Resonances

The response of the cell studied by Smoorenberg and Linschoten (1977) to the closely packed components above the sixth harmonic suggests a mechanism for encoding resonance frequency, whatever the characteristics of the source. The amplitude of the damped sinusoidal response of the cell was approximately proportional to the stimulus amplitude, and the response rose to its maximum amplitude when the stimulus energy had a maximum at the cell's characteristic frequency. Presumably, an array of such cells having different characteristic frequencies would reflect the spectrum envelope of a filter excited by any low-frequency or other suitably rich source. Although we have no direct evidence—in the form, for example, of responses of large populations of cells—we have here, corroborating the results of Sachs and Young (1980) in the auditory nerve, some indication of how vowel-like sounds may be represented in the cochlear nucleus.

Resonance Detectors in the Periphery?

We lack direct evidence for cells that act as feature detectors for resonances at certain frequencies in, say, the cochlear nucleus. Cells have, however, been identified in the cochlear nucleus that respond in a variety of ways to tones of specific frequencies (Evans 1974). These are good candidates for the kind of resonance detectors we might expect to find. Unfortunately, no systematic studies using vowel-like filters excited by buzzes or noise have been performed at that level of the auditory system. The experiments reviewed above provide evidence that the spectrum envelopes of sounds are represented in the auditory nerve and in the cochlear nucleus. It would be of great interest to determine experimentally whether or not further processing, such as the detection of resonances, is carried out peripherally as well.

F-PATTERN DETECTORS

Even lacking direct evidence of what we may call resonance detectors in the periphery, we can nevertheless ask if detectors of entire spectrum envelopes or, more explicitly, F-patterns appear to exist in the central auditory system. Since auditory physiologists studying the central nervous system have generally used more or less artificial stimuli made up of sinusoids, simple clicks, or unfiltered noise, we have little in the way of direct evidence. A few studies are exceptions, however, and their results are more directly applicable to the theory of sound color.

One of the first of these was by Keidel of Erlangen University (1974). Studying the medial geniculate bodies in awake and freely moving cats, Keidel found cells that appeared to fire only to vowel stimuli. The responses of one of these cells, which he called *vowel detectors*, are shown in Figure 25. Also in the medial geniculate Keidel found cells that he calls *consonant detectors* and *transient detectors*. Keidel suggests that the cat has no "reason" for such structures on evolutionary grounds but that it learns to respond to the voices of its handlers. As was suggested in Chapter Two in the justification for Rule 1, vowels (and possibly consonant-like sounds) can be thought of as examples of a much broader class of sounds produced by weakly coupled source/filter systems. Such sounds include environmental sounds that, it could be argued, would be to the selective advantage of the cat (and many other animals) to be able to discriminate. Keidel's vowel detectors do not need to be treated as the memory traces of learned behavior. They seem, rather, to be exactly the F-pattern detectors we are seeking.

Recent studies provide some corroboration of the Keidel experiment. Langner, Bonke, and Scheich (1981) found units in the forebrain of the mynah bird that increased their firing rate in response to certain vowels. The spontaneous-discharge rates of these same cells

FIGURE 25: *A cell in the cat medial geniculate that responds to the vowel [aa]. The upper trace is the neural response in coordination with the stimulus in the bottom trace. (From Keidel 1974.)*

were decreased by other vowels. The response of these cells to stimuli made up of tones swept across a broad range of frequencies revealed two separate frequencies that produced excitatory responses; they appeared, in other words, to be tuned to a particular combination of the first two formant frequencies. The pattern of response by these cells to single-formant sounds was predictable from the two-formant responses: excitation within the tuned range, suppression outside that range. These cells are good candidates for the F-pattern detectors we are seeking because they respond more strongly to the two-formant stimuli than to either F_1 or F_2 alone. The patterns of excitation and suppression of four separate cells are shown on the F_1–F_2 plane in Figure 26.

Suga, Kuzirai, and O'Neill (1981) have located areas on the auditory cortex of echo-locating bats on which are spread in two dimensions the frequencies of various components of the bat's own emitted orientation sounds.[6] They interpret these findings as providing an analogy in the bat to the F_1–F_2 plane upon which vowels must be distinguished by human beings. In other words, bats may not be as sensitive to sound color as other animals because the processing capabilities that would otherwise be devoted to resonance frequency analysis are taken up by the "analogous" echo-location system. However, if there is reason to regard the F_1–F_2 plane as significant to certain ani-

---

[6] In their remarkable experiments, Suga and his coworkers have identified the physiological functions—one could almost say the perceptual functions—of nearly all the multiple auditory areas in the bat's cerebral cortex. Their studies provide support for the suggestion of Merzenich et al. (1977) that the parallel, cochleotopically organized pathways in mammalian auditory systems encode different aspects of sounds.

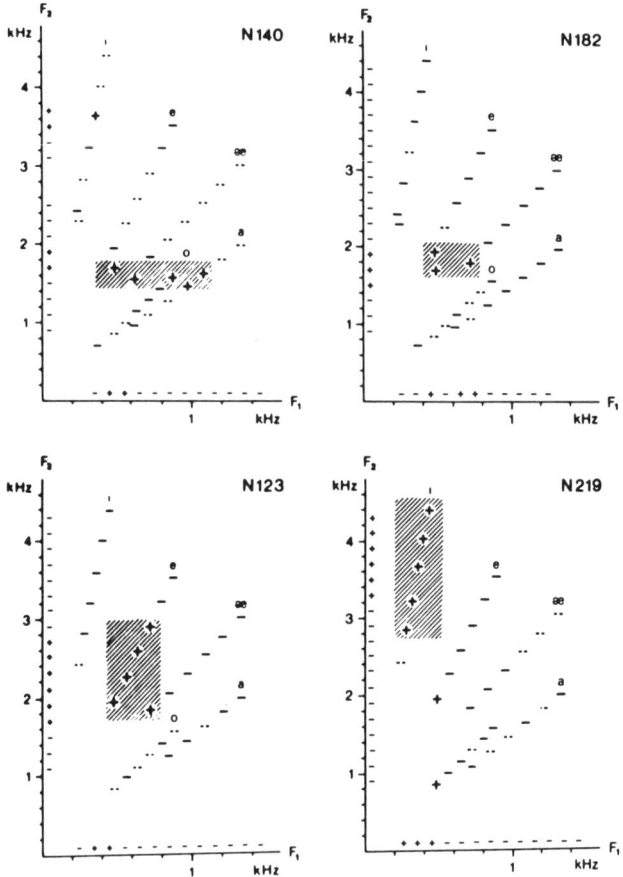

FIGURE 26: *Selective areas of cells in mynah bird forebrain. The pluses indicate excitatory responses; the minuses, suppression. F1 = abscissa; F2 = ordinate. Marks near the axes indicate responses to single formants. (From Langner et al. 1981.)*

mals other than man—in other words, if we can treat the space more generally as a sound-color space rather than simply a vowel space—then it makes sense to seek representations of it in experimental animals of those species.

DIFFERENTIATION OF SOURCE FROM FILTER

The studies reviewed above suggest that in the auditory system, structures on a number of levels appear to respond characteristically to the sonic correlates of resonances in filters. This suggestion supports the *invariance* part of Rule 1 in its three versions. Now to complete the argument, we must seek physiological correlates to the *variable* part of that hypothesis: we would like to be able to identify structures or modes of operation in the auditory nervous system that operate, *separately from the filter analyzers*, on the properties of the sound that Rule 1 permits to vary freely—those originating in the filter's source of excitation.

The fact that the auditory nervous system "turns on" with the onset of a sound is a quite general and obvious, but by no means trivial, example of such a mode of operation. In the discussion of ideal sources in Chapter Two, it was shown that the impulse is an adequate source for exciting a filter. The physiological response to the impulse is quite different from the detailed representation of the spectrum envelope imposed by the filter. The occasion for the impulse—its "date"—is simply signalled by abrupt changes, either increased excitation or suppression of spontaneous firing rates, in (effectively) all the neurons of the auditory nerve and in many of those in the higher levels of the auditory system (Whitfield 1967). The harder case—and one with which physiologists have been concerned nearly to the exclusion of all other aspects of sound—is that of a pitched source.

The Physiological Correlates of Pitch

A persistent problem in auditory physiology has been to account for the perception of pitch (Whitfield 1967, 142). The great precision with which human beings and the other higher animals are able to discriminate between sounds of different frequencies is the central difficulty; a mechanism must be found that is capable of making extremely fine distinctions in frequency. But frequency differentiation alone does not distinguish between filter and source characteristics. Both are coded in the frequencies of sounds.

The pitches of periodic sounds that have no energy at the fundamental are a different matter. This "case of the missing fundamental" (Stevens and Davis 1938) arises in sounds that are made up of a mixture of two or more tones whose frequencies are some integral multiple, greater than one, of a "missing" low frequency, F_0. The pitch of such sounds, as indicated by a number of different psychoacoustic methods (Schouten 1938, 1940; Houtsma and Goldstein 1972), is the same as that of a tone at the frequency of F_0. The term *periodicity pitch* (or *residue pitch*) for the pitch sensation is appropriate, because the sound is periodic at a frequency at which it contains no energy.

Nothing that we have seen in the response of the auditory system to filter characteristics would lead us to expect a phenomenon like this. No "missing formant" is heard when the resonances happen to fall into a ratio of, say, three to two. The discrimination of residue pitch must be a distinctly different process from the analysis of filters. If we can identify auditory mechanisms that appear to respond at the frequency of the fundamental of "missing fundamental" sounds, we shall have established a physiological basis for the analysis of an important kind of source: periodic, complex signals. And that physiological basis will be distinct from the physiological underpinnings of sound color. In contrast to the F-pattern detectors, the evidence for which was reviewed above, we now must seek *periodicity detectors*.

Periodicity Detectors

M. H. Goldstein et al. (1971) found a population of cortical cells in adult cats that they called *lockers*. These cells were observed to fire precisely in synchrony with a series of periodic clicks; they "locked" on to the fundamental period of the click train. Each of these cells had characteristic limiting rates above which, after an initial response at the onset of the click train, firing was suppressed. The locking behavior of these cells was preserved when the stimulus was replaced by periodic bursts of noise.

Møller (1970) has reported cells that respond like lockers in the cochlear nucleus. Britt (1976) has studied the details of the synaptic events of such cells. Similar cells that respond to the periodicity of stimuli have been found in the inferior colliculus and medial geniculate levels (Erulkar 1975). Whitfield (1980) argues in an interesting way from the results of his experiments with decorticate cats that true pitch perception requires the auditory cortex. He also suggests that the cortex is necessary—and, by implication, responsible—for perceptual constancies in general.

Specific studies of "missing fundamental" sounds have resulted in convincing demonstrations of physiological correlates of the periodicity pitch of those sounds. Smith et al. (1978) measured the so-called frequency-following response (FFR) in human listeners.[7] The spectrum of this signal in response to a tone at a given frequency, say F_0, appears very much the same as the response to a mixture of four tones at $2F_0$, $3F_0$, $4F_0$, and $5F_0$. The response to the single tone at F_0 can be reduced considerably by a narrow band of noise centered around F_0, but the same band of noise does not affect the FFR to the four-tone mixture. An extensive series of studies by Greenberg (1980) of the human FFR to synthetic vowels shows strong responses to the fundamental frequency and weaker responses to the first formant when the fundamental component is subtracted. The FFR exhibits little or no corresponding response to the second and higher resonances; it appears to be a measure of the auditory system's response to the periodicity of a sound rather than to the aspects of the sound that reflect filter characteristics.

A number of other studies that have identified physiological correlates of the missing fundamental have been reviewed by Evans (1978). The mechanisms for periodicity pitch detection remain unidentified, although some recent theories appear to be quite successful in accounting for both psychoacoustic and physiological results (Goldstein 1978). Some of these theories have attempted to incorporate musical phenomena (Terhardt 1977).

Even though we do not completely understand the auditory mechanisms by which periodicity pitch and other source characteristics are analyzed and those by which F-patterns are detected, the physiological evidence suggests both that such mechanisms exist and that they differ distinctly in character. This evidence, taken as a whole, appears to give considerable support to Rule 1 of the theory of sound color.

[7] The FFR is a signal, measured by gross electrodes on the surface of the skin around the head, that is thought to be made up of the massed firings of many auditory neurons, probably in the inferior colliculae.

DIMENSIONS AND OPERATIONS

In some of the discussion in this chapter we have tacitly assumed that resonance detectors are simpler kinds of feature detectors than F-pattern detectors. The analysis of single resonances is treated as a lower function that provides input to the analysis of patterns of resonances. But there is another possibility. Detectors of the central frequencies of the first two resonances could also be regarded as detectors of the values of a sound color on the dimensions of OPENNESS and ACUTENESS. This view is consistent with Rule 2 of the theory of sound color, which asserts that those two dimensions are closely correlated with the frequencies of F_1 and F_2 respectively. Information about the individual resonances could be carried to higher levels of the auditory nervous system along with information about the entire F-pattern. The direct evidence for resonance detectors seems less good than that for F-pattern detectors, although there are certainly candidates for such functions among the structures studied in the cochlear nucleus (Britt and Starr 1976).

A detail in the results of Langner, Bonke, and Scheich (1981) provides a suggestion of direct physiological support for the dimensions of ACUTENESS and OPENNESS. The selective areas of cells in the mynah bird's forebrain have shapes that suggest correlations with the frequencies of single formants (see shaded areas in Figure 26). The cells appear to respond only to narrow frequency ranges of one of the first two formants while responding over a wider range of frequencies to the other. An array of these cells having the same value for one resonance and different values for the other might provide a basis for measurement along a dimension.

Aside from that single study we have no direct physiological evidence for the dimensions of sound color. It is appropriate to emphasize that the paucity of evidence must be viewed in light of the absence of attempts on the part of physiologists to seek physiological correlates of the sound-color dimensions. With the definition in the present study of the sound-color dimensions and the suggestion that sound color may be an aspect of auditory sensation in many of the higher animals, physiological study of the dimensions seems appropriate and plausible.

OPEN QUESTIONS: A MODEL AND SOME PROPOSED PHYSIOLOGICAL RESEARCH

In the course of an extended discussion of their results on the representation of vowel spectra in the auditory nerve, Young and Sachs (1979) conclude that the identification of periodicity pitch requires the same kinds of information as identification of vowels: some kind of representation of the spectrum of the signal. They point out that the frequencies of the partials in that spectrum determine the pitch, whereas the frequencies at which the spectrum representation rises to maxima determine the vowel.

Let us attempt to analyze what kinds of mechanisms vowel, or sound-color, detection and pitch detection require. Figure 27 presents, in functional outline, models for F-pattern detection and for periodicity detection. Although the periodicity detector in this scheme is unrealistic on several grounds, including its failure to recognize such salient musical phenomena as octave equivalence and the equivalence of other musical intervals (perfect fourths sound like perfect fourths at whatever absolute pitch they may be played), even this simple pitch-model contrasts with the F-pattern detector model. The two are different in the requirements they impose on the bandwidths and the frequencies of the peripheral analyzer that supplies their inputs. The F-pattern detector is less complex; it could be a simple cross-correlation device, for example, and it requires only moderate resolution in frequency.[8] The pitch detector, on the other hand, requires a kind of locking on to specific narrow bands that are at integral multiples of the fundamental. With two such contrasting analyses of the frequencies in sounds, a number of interesting questions arise that revolve around the separation of source and filter characteristics in the responses of the auditory system.

The first such question concerns the representation in the auditory nerve fibers of the spectrum envelope of a vowel when the excita-

---

[8]C. L. Searle (1982) discusses the specifics of a model similar to the one discussed here. I am grateful for the help of Searle and Richard Stern in formulating this model.

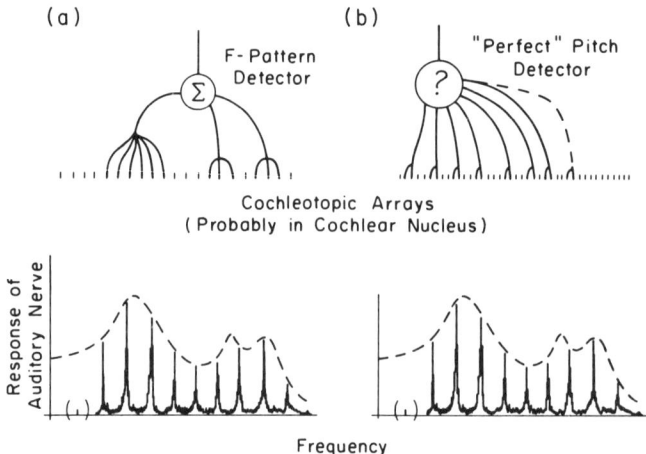

FIGURE 27: *Models of F-pattern and periodicity detectors. The F-pattern-detector neuron receives input from specific broad-frequency regions in (a). The perfect-pitch detector also takes input from different frequency regions, but from very narrow bands in (b). The pitch detector is locked into frequency bands that are in integral multiples of the fundamental, whereas input to the F-pattern detector is from arbitrary frequency bands. The "missing fundamental" does not affect either detector appreciably.*

tion is other than a low-frequency buzz. What would the methods of Sachs and Young show if a frequency-modulated source were used? It would also be interesting to see how robust the auditory nerve representation would be when the source is at a high frequency—a source, in other words, that samples the spectrum envelope only sparsely.

Even more revealing would be systematic investigation of the hypothetical feature detectors, both F-pattern and periodicity, for which we have only suggestive evidence. Do the cells found by Langner, Bonke, and Scheich (1981) in the mynah bird respond in the same way when the source and the filter of the synthetic vowels are arbitrarily related? Will experiments with other species show cells with characteristics like those in the mynah bird? Are the responses of locker cells affected by a strongly characteristic resonance structure imposed on the partials of a periodic source?

Given the recent interest in speech sounds as stimuli, it is likely that physiologists are exploring these and similar questions at the present time. The generality of the source/filter model of sound suggests that studies of the independence of source and filter characteristics are appropriate, whether or not a separate rationalization can be found to study speech per se. It seems appropriate and timely to engage in such studies across a broad range of species.

CHAPTER FOUR

Evidence from Psychoacoustics

THE FIRST TWO postulates of the theory of sound color, Rules 1 and 2, are claims about the relationships between certain parameters of sound and listeners' perceptions. It is precisely these kinds of relationships that psychoacousticians take as their field of study, so it is particularly appropriate to ask what support there is from that field for the theory. Through psychoacoustic experimentation, not only can the theoretical claims be tested directly, they can be cast, potentially, into more specific and—to some extent—quantitative form.

LIMITATIONS OF PSYCHOACOUSTIC METHODS

In contrast to most physiological methods, psychoacoustics makes it possible, in principle, to study the operation of the normal human auditory system as a whole. Human listeners can tell the experimenter what they hear directly. But the methods of psychoacoustics introduce certain limitations of their own. Indeed, the capabilities and limitations of psychophysical methods[1] are a study in and of itself (S. S.

---

[1] Psychophysics is the study of the relationships between physical measures of stimuli and measures of sensation in all sensory modalities—vision, taste, etc.; psychoacoustics is the psychophysics of hearing.

Stevens 1951, 1958, 1971; Swets 1961; Luce and Galanter 1963; Shepard 1966, 1974; Marks 1974). There has always been controversy regarding certain methods. However, within the last two decades or so, broad areas of agreement have emerged, and although some methodological issues are not yet settled, the solution to many psychophysical problems can now be sought with a degree of confidence in the validity of the results.

Threshold Measurements

Among the areas of general agreement about the limitations of psychophysical methods is the status of absolute or differential threshold measurements. *Absolute thresholds* define the limits of the sensory world of human beings and animals. *Differential thresholds* are measures of the sensitivity of the sensory system to changes in stimuli. Threshold measurements of both kinds are expressed in physical quantities—the luminance and wavelength of a light that can be just discriminated from darkness or from another light; the temperature, shape, and location of an object in contact with the skin that just gives rise to a sensation of warmth; or the sound-pressure level and frequency spectrum of a sound that can be heard as just different from another sound or from silence.

Even though threshold measurements are expressed in physical terms, they are by no means exact. It is impossible, for example, to specify the sound-pressure level of the softest audible sound. We can specify a sound-pressure level at which young adults with normal hearing, in a certain experimental setting, report that they can hear a 1000 Hz tone 75 percent of the time. But if the criterion were shifted to a 50 percent "hit" rate, a different value for the absolute threshold would result. A number of standard methods of determining the absolute threshold have been used, each giving a slightly different result. We can never measure *absolutely* the absolute threshold in any sensory modality (Swets 1961).

The differential threshold—sometimes called the *difference limen* (DL) or *just noticeable difference* (JND)—is also defined probabalistically. Several methods have been used, and each involves certain arbitrary choices—e.g., a criterion of 75 percent or 50 percent correct responses—like those made in absolute-threshold studies (Marks 1974). Thus although the size of the JND is quantified as a certain difference in frequency, intensity, etc., that number is only a statistic that is dependent on the details of the experiment by which it was measured.

A theory of what have come to be known as *detectability* experiments has been developed in the last twenty years. The theory, which attempts to deal consistently with threshold measurements, raises hopes of identifying the sources of variability in those measurements (Luce and Galanter 1963; Green 1976).

Measures of Psychological Distance

Among the oldest and most intriguing problems in psychophysics is to determine the *scale* relating distance along physical dimensions to distance in the psychological or sensory domain.[2] The most prominent and successful attempts to solve this kind of problem are those of S. S. Stevens. He developed a set of so-called direct methods: magnitude estimation, magnitude production, and cross-modality matching. In magnitude estimation, observers provide measures of psychological distance by estimating numerically the sensory magnitude of some aspect of a given stimulus. In magnitude production, the observer sets a stimulus parameter—usually by turning a knob—in an attempt to match sensory magnitudes with numbers given by the experimenter. In cross-modality matching, the observers manipulate a device that permits them to set the magnitude of one sensory modality—say, finger pressure on a key—so as to match the perceived magnitude in another modality—say, brightness of a light. Issues that arise in using these scaling methods are discussed by Stevens (1971), by Marks (1974, especially chapter 7) and, with some disapproval, by Luce (1972).

The question of the validity of a psychological scale is complex (Marks 1974, 30–31, chapter 7). It cannot be decided, for example, by

---

[2]Marks (1974) provides an excellent historical review of the attempts to solve these kinds of problems.

simple estimates of the amount of variability in a particular experiment. The "direct" methods seem valid—that is to say, repeatable, in reasonable agreement among themselves, and so forth—when they are used to scale single perceptual dimensions such as loudness. Unfortunately, they sometimes break down in complex situations (Marks 1974, 57–58), a fact that calls for caution in applying them to the study of the multidimensional space of sound color.

Some techniques have recently been developed to deal specifically with multidimensional perceptual phenomena. An excellent review of these methods has been published by one of their leading proponents, Roger Shepard (1980).

The Problem of Context

In an effort to control as many variables as possible, psychophysicists usually present stimuli in a neutral or isolated context. This neutralization of context raises the question of whether the results of the experiment can be applied to everyday listening or to a musical context. It is possible to distinguish two kinds of context effects. In the first the context can be said to change or mask the aspect of the stimulus that is under study. These effects can be studied directly, using psychoacoustic methods in which the context becomes one of the variables controlled in the experiment (e.g., Grey 1978).

The second kind of context effect is more difficult to deal with. In this case the perception of a sound is said to be altered depending on whether or not one is listening in a special "mode." The "speech mode," which has received a great deal of attention, will be discussed in Chapter Five. Here it is enough to observe that there may be a number of such modes, possibly including a "music mode," and that psychoacoustic results must always be interpreted tentatively until they are replicated in contexts controlled for such modes of listening.

The Problem of Learning

Related to the problem of context in a psychoacoustic experiment is the question of whether one is studying the biological features of the auditory system, or a learned response. The effects of learning are presumed, often tacitly, to be eliminated by studying the responses of numbers of listeners whose experiences, although contributing to the variability of the experimental results, leave unaffected the mean values—i.e., the underlying biological property of the auditory system. Needless to say, this presumption is a dangerous one. There is really no way of knowing the extent to which learning may have affected the results of any single experiment. Only repeated studies, using a variety of different methods that produce the same results, permit us to conclude that what we are studying is likely to depend largely on biological structures.

PSYCHOACOUSTICS OF FILTER SYSTEMS

Psychoacousticians, like auditory physiologists, have largely neglected the study of the perception of filter characteristics. Sounds that are the result of a filter system acting on a source are considered by many researchers to be too complex for productive research. Others have attempted, with some success, to apply psychoacoustic techniques to the study of the perception of resonances, F-patterns, and spectrum envelopes. One of the most fundamental questions asked by these scientists has to do with our sensitivity to changes in resonance frequency.

Sensitivity to Changes in Filter Parameters

In an experiment using damped sinusoids as a stimulus, K. N. Stevens (1952) measured the sensitivity of listeners to changes in the "pitch" of single resonances. The difference limens (DLs) he obtained varied as a function of both the frequency and the bandwidth of the resonance. His results, at resonance bandwidths typical of those found in spoken vowels,[3] indicate that the DL is about 2 or 3 percent of the resonance frequency. In other words, the experimental results would lead us to expect that changes of 10 or 15 Hz in the fre-

---

[3] Specifically, 50 Hz for resonance frequencies below 500 Hz, and 10 percent of the resonance frequency for higher frequencies.

quency of a resonance centered at 500 Hz would be detected by listeners 75 percent of the time.

Flanagan (1955) studied sensitivity to changes in the formant frequencies of synthetic vowels and found the DL to be about 3 to 5 percent of the resonance frequency. This value is about ten times the DL he found for changes in the fundamental frequency of the same kinds of sounds (Flanagan and Saslow 1958).[4] He concluded on the basis of later experiments (1965) that the DL for the bandwidths of resonances in synthetic vowels is between 20 percent and 40 percent—or about ten times the DL for resonance frequency.

Kakusho, Kato, and Kobayashi (1968) repeated a portion of Flanagan's experiment and, expressing their measurements in a different way, found a greater sensitivity to formant frequency change than had Flanagan. Their "just discriminable change" (JDC) for F_1 and F_2 was 0.9 percent of the resonance frequency.

In an interesting set of experiments, Haggard (1977) showed that splitting the formants—sounding the first formant in one ear and all the higher formants in the other ear—increases sensitivity to changes in formant frequency. Haggard interprets these results as indicating the degree to which the higher resonances are masked by the generally high amplitude of the first formant.

Given certain assumptions, we can compare the DL for formant frequency with the size of what is known as the *critical band*, the bandwidth of frequencies within which energy is said to be summed to make up the loudness of a complex sound (Zwicker and Scharf 1965). The width of the critical band varies with frequency, and it is closely proportional to the DL for changes in formant frequencies within the range measured by Flanagan (1955). However, in absolute terms, the critical band is approximately three times the size of the formant-frequency DL (Scharf 1970).

---

[4]This result is in essential agreement with early studies of the JND for changes in frequency of sine tones (Shower and Biddulph 1931) and a study of the repetition rate of a train of periodic pulses exciting a bandpass filter with the lowest eight or nine partials filtered out (Ritsma and Hoekstra 1974). Values of 0.3 to 0.5 percent change in frequency were obtained in all these cases.

Interpretation of the Sensitivity Measures

K. N. Stevens's experiment with single resonances eliminated any suppression that resonances in multiple-resonance systems might impose on each other, so his results can be taken as setting a minimum for the value of the DL for resonance frequency. Flanagan (1965), in deriving his estimate of the same statistic, lumped together tests of three different standard F-patterns from which deviations were judged. In those of Flanagan's F-patterns that most closely matched the single pattern tested by Kakusho, Kato, and Kobayashi (1968), the DL was somewhat smaller than in the others he studied. Kakusho, Kato, and Kobayashi found that noise excitation resulted in a larger "matching range"—a measure similar to their JDC—of 1.5 and 2.1 percent for F_1 and F_2, respectively. They suggest that their results are actually in agreement with Flanagan's, because some earlier DL measurements of their own yielded values two to five times larger than their JDC. This explanation for the divergence from Flanagan's results is not entirely satisfactory, because the JDC measures for fundamental frequency by Kakusho, Kato, and Kobayashi are in agreement with Flanagan and Saslow's (1958) DL measure for that parameter.

The stimuli of both Flanagan (1955) and Kakusho, Kato, and Kobayashi (1968) were synthesized in ways that made them sound somewhat artificial, and the stimuli of both K. N. Stevens (1952) and Haggard (1977) were intentionally artificial. This unnaturalness in the stimuli may account for the remaining rather small discrepancies among the results of the experiments. The critical band[5] is considerably larger than the DLs obtained directly, but its close proportionality to the DL suggests a possible relation between the two measures.

Having at least an approximate measure of sensitivity gives us some basis for calculating the number of different sound colors that can be distinguished within a given F_1–F_2 area. Let us assume Flanagan's (1955) largest value for the DL of resonance frequency—5 per-

---

[5]This measure has been compared (Pick 1977) to the physiological measures of "grating acuity" by Smoorenburg and Linschoten (1977) and by Evans (1977) that were reviewed in Chapter Three.

cent—and assume further that the DL indicates something about the number of different sound colors that can be discriminated over a range of resonance-frequency settings. It would follow that 20 sound colors can be distinguished within a range defined by doubling one of the resonance frequencies. If we take the area on the F_1–F_2 plane between 400 Hz and 800 Hz for F_1 and 1000 Hz and 2000 Hz for F_2, we could conclude that 400 sound colors can be distinguished within that area alone! Even if the critical band is taken as a better indicator for this kind of calculation, our estimate would be reduced only to 45 discriminable colors, still a large number within a restricted part of the F_1–F_2 plane.

By no means can these estimates be used as indications of the number of sound colors that can be recognized reliably in an arbitrarily complex context. Moreover, since the DL is a statistical quantity related to the discriminability of a change in resonance frequency, it cannot properly be treated as if it directly measured the number of discriminable entities. Nevertheless, these studies suggest that, under ideal conditions—conditions, incidentally, that may be met in certain special musical contexts—a very large number of sound colors are distinguishable within the "normal" sound-color space.

Mechanisms for Resonance Detection

Among Békésy's ingenious studies of the actions of cochlea-like mechanisms is his demonstration that the skin integrates the sensations from multiple vibratory stimuli (1957; 1960, 590–609). He placed five spatially separated vibrators, tuned successively to vibration rates an octave apart and equal in amplitude, simultaneously on the skin of an observer's arm. The sensation felt was a single maximum localized at the center vibrator with a vibratory "pitch" equal to that of the center vibrator when it was presented alone (see Figure 28). Increasing the amplitude of one of the vibrators moved the apparent location and the vibratory "pitch" toward that vibrator. Employing the vibrators as a model of the mechanical action of the basilar membrane and the skin senses as a model of the auditory nerve, Békésy interpreted his findings as evidence for "lateral inhibition," a neural mechanism he postulated

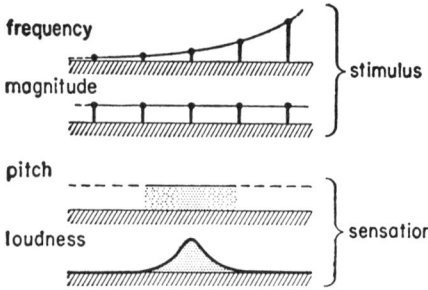

FIGURE 28: *Vibrators on the skin: a model illustrating "lateral inhibition" in the auditory system. (From Békésy 1960.)*

to account for the sharpening of the flat peaks in the mechanical motions of the basilar membrane. His interpretation has been criticized on the grounds that the fine-tuning of the auditory nerve fibers acts too fast to permit neural inhibitory mechanisms to take effect (Evans 1974). However, the psychophysical results themselves suggest the presence of some mechanism that integrates across space (hence, in the ear, across frequency) to produce a single sensation from multiple, separate stimuli.

In another series of experiments with vibrators on the skin, Békésy applied random continuous signals alternately to two vibrators with a switching rate of from 0.25 to 0.5 seconds. The sensation was felt to be centered between the vibrators "when the distance between the [vibrators] was small" but was localized to one or the other as the separation was increased (Békésy 1960, 633–34). This "funneling," as Békésy called it, of the separate vibratory stimuli into a single sensation *localized between the two vibrators* provides a possible model for resonance "localization" between two partials in a sound produced by a periodic source exciting a filter system. This second experiment of Békésy's lends support to the idea of a genuine resonance detector in the periphery rather than a mechanism that depends on the ear treating the most intense partial as a resonance.

Studies by a group of researchers in Leningrad led by Ludmilla Chistovich have provided interesting corroborative evidence. Chistovich (1971) has interpreted the results of an experiment with synthetic vowels having unusually high fundamental frequencies as support for the most-intense-partial option. This appears to agree with Békésy's results in the case of very widely spaced vibrators. Békésy's "funneling," which appears with closely spaced vibrators, parallels what the Leningrad group call the "center of gravity" effect—a single resonance substituting for two resonances. Chistovich and Lublinskaya (1979), studying the perception of one- and two-formant vowels in which the formants are closely spaced, showed that a single formant can be matched to a two-formant vowel provided that the two formants are within about 3 to 3.5 critical bands. When the formants are more widely separated, the "center of gravity" effect breaks down and even small spectral peaks are treated as formants.

In comparing these studies to the physiological results reviewed in Chapter Three, we are struck by the importance of investigating the independent effects of filter characteristics—the "funneling" of multiple partials into a single sensation—and the effects of the source—the spacing of the partials. The results reported by Sachs and Young (1980) can be interpreted as suggesting that the "funneling" takes place above the level of the auditory nerve—an argument against Békésy's view that it operates in the cochlea. The indications of F-pattern detectors in the thalamus (Keidel 1974), on the other hand, suggest that the spatial integration has been accomplished at or below that level. Systematic, independent variation of source and filter in these physiological studies is required to identify and locate possible feature detectors of resonances.

Psychoacoustic Scales for Resonance Frequency

Given a change in the frequency of one of a filter's resonances, how much does the sound color change? The function relating distance along a perceptual scale of sound color to changes in F-patterns has yet to be determined. A number of technical difficulties stand in the way of that determination, of which the multidimensionality of color is probably the most critical. The usual methods of dealing with a multidimensional space involve asking observers to make similarity or difference judgments among pairs of stimuli within the space, but ordinarily these judgments cannot readily be interpreted as determining psychological distance measures directly.[6]

Bismarck (1974b) attempted to scale the "sharpness" of a series of synthesized complex sounds having a variety of spectrum-envelope slopes, high-frequency cut-offs, and types of excitation (periodic, noise, etc.). Using methods that involved asking the listener to make doubling and halving judgments, Bismarck demonstrated that sharpness judgments could be made by listeners, but he could only conclude that "sharpness appears to be primarily related to the position of the loudness concentration on a critical band-rate scale rather than to a particular shape of the spectral envelope" (171). Bismarck's use of stimuli that were patently "synthetic" makes it hard to interpret his results in terms of resonance perception, but the reference to "position of the loudness concentration" suggests an analogy to the effect of a resonance. Well-known biases introduced by the experimental method (Marks 1974, 16)—especially the interpretation of a sensitivity measure as an indicator of position on a scale—weaken the effect of the data in determining a psychophysical relation. Nevertheless, Bismarck's suggestion that sharpness may vary with the critical band (or Bark scale: see Zwicker and Feldtkeller 1967) and the proportionality of that scale with the DL for changes in resonance frequency suggest in turn that his study may provide a beginning for the determination of a distance measure in sound-color space.

The lack of an established psychophysical relation between sound color and resonance frequency precludes further precision in the assertions of the theory of sound color. If the mel scale (see below), the critical band scale, or some other function could be established as the proper scale relating resonance frequency to sound-color distance, the acoustic interpretation of those rules could be made explicit. In the absence of such a scale, the theory must be stated rather loosely—for example, Rule 2 merely referred to figures on which equal-value contours were drawn approximately. It is worth emphasizing once again, however, that determination of more exact quantitative relations be-

---

[6] In certain cases some indication of a psychophysical scale can be estimated from the results of calculations on such judgments (Shepard 1980).

tween F-patterns and sound color is unlikely to require more than adjustments of the equal-value contours. Other aspects of the theory—in particular, the operations on sound color—would probably not have to be significantly altered.

PSYCHOACOUSTICS OF SOURCE CHARACTERISTICS

The fundamental postulate of the theory of sound color is that filter characteristics are perceived separately from source characteristics. Psychoacoustics should provide clear evidence for that separation. We have reviewed some studies of filter characteristics. What about psychoacoustic studies of the source?

We need not cite evidence for the fact that people hear certain sources. Our everyday experience tells us that the onsets of sounds, and the accompanying general activity in the auditory nerve, get our attention. By analogy to Chapter Three, we need not inquire into the perception of single pulses, which we may take to be a representative kind of "event-like" sound; the fact that we hear them is enough.[7]

Pitched sounds are another matter. Because the properties of both source and filter in sounds having pitch are carried by frequency, there is a prima facie reason to suppose that the two may be confounded—contradicting Rule 1 of the theory of sound color. It is important, then, to compare studies of the perception of periodic sources to studies of the perception of resonances.

Registral Pitch: The Mel Scale

Stevens, Volkmann, and Newman (1937) and Stevens and Volkmann (1940) established a scale relating perceived pitch to frequency that they called the mel scale. The scale, neither strictly logarithmic—as

---

[7]This is not to say that pulses and pulse trains are not interesting stimuli for psychoacousticians to investigate. Quite the contrary, much insight into hearing processes has been derived from experiments with exactly those sounds (reviewed in Flanagan 1965). However, those investigations are not concerned with pulses as *sources*—as indicators, that is, of the occurrence of some mechanical action in the environment.

one might expect from the musical-pitch scale—nor linear, encompasses the entire range of audible frequencies. Interestingly, it correlates well with a variety of other psychophysical and some physiological measures, including the relation of frequency to the distance of the point of maximum excitation along the basilar membrane (Stevens and Davis 1938, 96). Among the methods used to derive the mel scale was the "fractionation" method, in which listeners were asked to set the frequency of one sinusoid to a pitch that was some prescribed fraction of the pitch of another standard tone. The tone under the listener's control was to be set to one-third of the pitch of the standard, for example. Musicians, used to thinking in terms of octave equivalence and musical intervals, are usually hard-pressed to imagine how they might respond to such a task; however, the results from the few musicians who participated in the experiments did not differ significantly from those of the non-musicians.

The mel scale was said (Stevens, Volkmann, and Newman 1937) to pertain to one aspect of musical pitch, its *height*, as distinguished from its pitch class or *chroma*. The distinction between these two aspects of pitch is common in Western music, and it seems to hold cross-culturally (Deutsch 1978). It is perhaps clearest in the theory of atonal and dodecaphonic music. The 12-tone row or a smaller pitch-class set controls the chroma, whereas the height—the register—is chosen according to other considerations. A provocative and difficult question is whether composers' registral choices have in some fashion reflected the mel scale instead of the more obvious log-frequency scale. Attneave and Olson (1971) found that, when listeners were asked to produce transposed versions of given melodies, the transposition was a log-frequency transformation of the original. The frequency ranges covered in the experiment coincided with a section of the mel scale that is nearly logarithmic, however; in addition, the experimental design did not rule out the possibility that the listeners were influenced by pitch class in the transposition task. The experiment, therefore, only weakly discredits the mel scale as a measure of perceptual distance in pitch.

An alternative interpretation of the mel scale follows from the possibility that listeners in the original mel-scale experiments were not judging a dimension of *ordinary* auditory sensation at all. Sine waves, after all, do not occur in nature, and the simplicity of their mathematical specification is not reflected in the sensations to which they give rise. Listeners in the mel-scale experiments may have been forced by the poverty of the acoustical stimulus to listen in a "reduced" manner. When faced with a sound that had neither genuine pitch nor color, listeners gave responses that, in effect, directly reflected some measure of distance along the basilar membrane and not the higher levels of auditory analysis that must underlie both pitch and color determinations in natural sounds. If this interpretation is correct, the mel scale pertains neither to the source nor to the filter but, rather, to an auditory process in the cochlea that underlies both pitch and sound color.

Periodicity or Interval Pitch

Music theorists have concentrated almost entirely on the organization of pitches as *chroma*, i.e., pitch and interval classes (Forte 1973). This kind of pitch may be closely related to what auditory physiologists and psychoacousticians have called *periodicity pitch* (Schouten 1970). The term *interval pitch* has been advocated as a synonym to suggest the musical character of this kind of pitch more explicitly (Slawson 1975). Although there has long been an interest in this aspect of pitch (e.g., Seebeck 1841; Stumpf 1890), the first extensive work on the subject, using modern methods of controlling the stimuli, was undertaken by Schouten (1938, 1940, 1970).

In a summary of his observations, Schouten (1970) concluded that in "residue pitch" (a phenomenon closely related to periodicity pitch):

> 1. The ear can distinguish tones whose frequencies are wider apart than a full tone. In a harmonic series some eight to ten lower harmonics can be perceived by the unaided ear.
>
> 2. Higher harmonics are heard collectively as *one* subjective component (one percept) called the *residue*.
>
> 3. The residue has a sharp timbre.

4. The residue has a pitch equal to that of the fundamental tone. If both (i.e., the residue pitch and the fundamental) are present in one sound, they can be distinguished by their timbre.

Not only can a pitch be heard in the "case of the missing fundamental," but according to Schouten, that pitch is unaffected if the first ten partial are missing as well. These and other observations led Schouten to conclude that some kind of analysis, not of the spacing of the harmonics or of the time envelope of the waveform, but of the periodicity of the sound, gives rise to the pitch of the residue (1970). Licklider (1954), Boer (1956), and Ritsma (1962, 1963) have replicated and largely confirmed Schouten's observations.

Recent studies have cast more light on the phenomenon of periodicity pitch. The residue pitch has been found to depend more strongly on the resolvable partials than on the unanalyzable components of a sound—Schouten's original residue (Boer 1976). Moreover, a number of experiments have begun to suggest that pitch perception is a function of the central nervous system. Houtsma and Goldstein (1972) presented listeners with two upper harmonics of a series of notes of a simple melody. The ability of listeners to identify the melodies was essentially unchanged whether the harmonics were both presented to one ear or one in each ear. In agreement with the physiological studies reviewed in Chapter Three, the weight of psychoacoustic evidence suggests that this kind of pitch sensation is the result of at least a two-stage process: some kind of frequency analysis, probably in the peripheral auditory system, and some central mechanism that operates on the results of that analysis (Boer 1977).

PSYCHOACOUSTIC COMPARISONS OF SOURCE AND FILTER CHARACTERISTICS

Only a few psychoacoustic studies have directly compared the perception of the source and the perception of resonances and F-patterns. These studies provide the most critical tests of Rule 1 of the theory of sound color.

A series of experiments with synthetic speech-like sounds (Slawson 1968) involved systematic variation of filter parameters with different frequencies of periodic sources. In two of those experiments, listeners were asked to estimate the difference in either "instrumental color or timbre"[8] or vowel quality between pairs of sounds presented in a series. The first member of each pair was one of six standard sounds and the second member of the pair was a modified version of the first: the fundamental frequency of the second sound was always shifted by a factor of 1.5 (up by the interval of a fifth) and the resonance frequencies of the second sound were shifted by one of eight factors ranging from 0.9 to 1.6. The smallest differences in both instrumental color or timbre and vowel quality were obtained when the resonances were kept constant in frequency (a formant-shift factor of 1.0) in the sounds with the higher fundamental or source frequency. As the resonances were shifted above 1.0 by increasing amounts, the differences reported by the listeners increased monotonically (see Figure 29).

The second-largest of the eight formant-shift factors—a value of 1.5—was equal to the shift of the fundamental. Pairs having that shift factor satisfied the conditions of the relative-pitch theory of timbre. However, no minimum in the difference estimates coincided with those pairs. The results clearly and unequivocally support Rule 1: the source frequency in the test sounds was raised, but given a range of possible F-patterns in those sounds, the listeners picked out the unchanged F-pattern as having the smallest differences in color.

These results were confirmed in an experiment by Plomp and Steeneken (1971). They used one-third-octave, band-pass-filtered pulse trains as stimuli and obtained dissimilarity judgments among nine stimuli: all combinations of three settings of the repetition rate of the pulses and three settings of the band-pass filter. Application of the MDSCAL program (Kruskal 1964) to the judgments suggested a two-

---

[8] Although the listeners in these experiments, and in some of those to be cited below, were asked to judge "timbre," their responses can be interpreted as pertaining to sound color whenever the stimuli resemble those produced in source/filter systems. Since sound color is to be regarded as a subset of timbre, the two terms can be considered synonymous and can be used interchangeably to instruct listeners, as long as the stimuli are suitably restricted in the experimental setting.

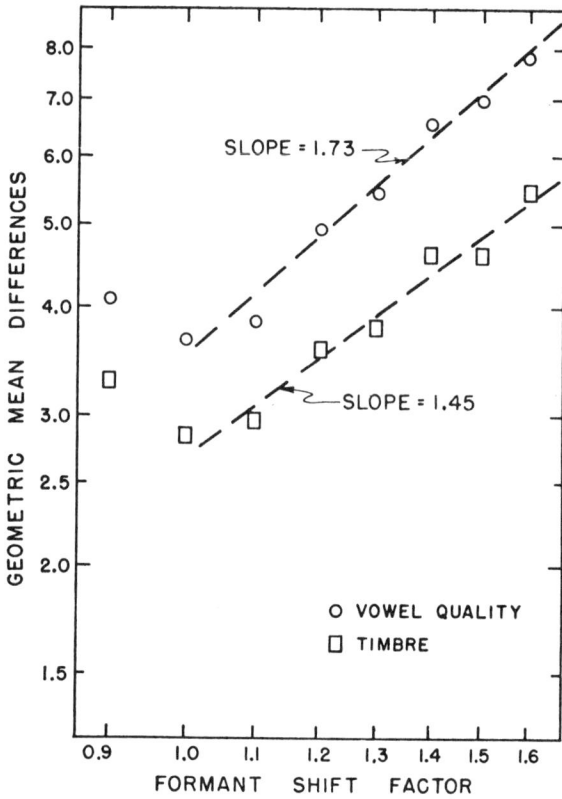

FIGURE 29: *Estimates of differences in vowel quality and timbre as a function of F-pattern shift factors in sounds with a fundamental frequency shifted by a factor of 1.5. (From Slawson 1968.)*

dimensional interpretation that showed clearly the two perceptual attributes, pitch and timbre (see Figure 30). The stimuli included both fixed-pitch (stimuli on the vertical lines, Figure 30) and relative-pitch transformations (stimuli 1, 5, and 9, Figure 30). The distances, and thus the observers' judgments of dissimilarity, between the stimuli with a constant spectrum envelope were smaller than those between stimuli with a constant harmonic content (diagonals with positive slope). It follows that the fixed-pitch theory of timbre—and its generalization in Rule 1 of the theory of sound color—was favored over the relative-pitch theory.

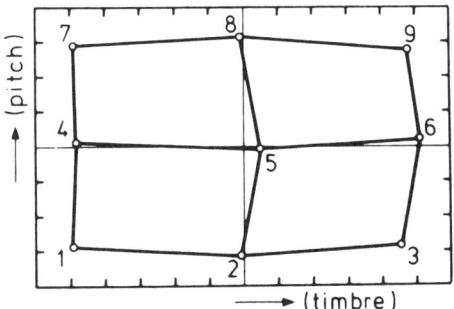

FIGURE 30: *Perceptual space of sounds differing in spectrum and fundamental frequency. The principal factors were timbre and pitch. (From Plomp and Steeneken 1971.)*

Cues for Normalization

The study cited above (Slawson 1968) also dealt with the effects of octave shifts in fundamental frequency. When the fundamental frequencies of the members of sound pairs were raised by a factor of two, a 10 percent shift upward in the resonance frequencies (a formant-shift factor of 1.1) resulted in the smallest differences in both vowel quality and musical timbre. This slight deviation from the invariance predicted by Rule 1 has been interpreted by Plomp (1976, 108) as a reflection of listeners' experiences with men, women, and children talking, and with other sounding objects, in which high-frequency sources are often associated with high average-resonance frequencies. According to this interpretation, listeners would have learned to use the high-frequency source as a cue that brought into effect the normalization process usually associated with "stretched" sound-color spaces. A demonstration by Ladefoged and Broadbent (1957) that vowel identifications can be affected by the average resonance frequencies in a frame sentence suggest another cue that evokes the process of normalization in sound color.[9]

---

[9] Further studies of normalization in the context of speech will be reviewed in Chapter Five.

PSYCHOACOUSTICS OF THE SOUND-COLOR DIMENSIONS

Little physiological evidence was cited in Chapter Three that pertained to the dimensions of sound color postulated in Rule 2 and the equal-value contours in Figure 13. Although the situation is not markedly better in the case of psychoacoustic research, a few studies have been carried out that pertain indirectly to the sound-color dimensions.

Dimensions of Complex Sounds

Among the earliest attempts to investigate the attributes of complex sounds was a study by W. H. Lichte (1941). Although marred by methodological problems, this study identified three dimensions, called *brightness*, *roughness*, and *fullness*. The first of these was found to be a function of the "midpoint of the energy distribution" in the sounds. Although the spectrum envelopes of Lichte's stimuli did not have the broad peaks characteristic of resonances, their slopes over the middle of the frequency range varied from that of a [uu]-like color to that of a strongly "stretched" [ae]. Lichte's *brightness*, in other words, parallels a conglomerate of the OPENNESS, ACUTENESS, and SMALLNESS dimensions postulated in Rule 2. Lichte gave no rationale for choosing the sounds he investigated, so his conclusions are hard to generalize. Moreover, his experimental method included an initial presentation of the extremes of his *brightness* dimension, a procedure that could have biased his results. Nevertheless, this study can be interpreted as early, rather equivocal, support for the multidimensionality of sound color.

Multidimensional Scaling Studies

The techniques of multidimensional scaling (MDS) and the semantic differential have been developed in the decades since World War II to investigate objectively relationships among stimuli that differ with respect to more than one psychological attribute or dimension. A few researchers have attempted to apply these techniques to the study of the timbre of complex sounds.

One of the earliest MDS studies of timbre was carried out by

Plomp (1970) with stimuli made up by selecting a single period from the sounds of nine different musical instruments, all played at the same pitch, and replicating that single-period waveform repeatedly to produce an extended pitched sound. The listeners were asked to select the most similar and the most dissimilar pairs from groups of three pairs. Measures of dissimilarity among all pairs of the nine sounds were derived from these judgments. Plomp later (1976, 93–97) analyzed these measures using the MDSCAL computer program. He obtained a geometric representation of the dissimilarity judgments, but rather than interpreting the three principal dimensions of perceived dissimilarity directly, he used the same MDS technique to analyze acoustic measures of the differences in the spectra of the stimuli. The resulting acoustic dimensions were found to correlate closely with those derived from the dissimilarity judgments, leading Plomp to conclude that musical-instrument timbre is strongly dependent on spectrum.

Plomp's three dimensions can be said to demonstrate the multidimensional nature of the sounds of musical instruments. Since Plomp omitted from his report the detailed acoustic specifications of his stimuli, it is hard to compare his findings to the predictions of Rule 2 and the accompanying equal-value contours. However, his first dimension, accounting for the largest proportion of the variance, distinguishes small, treble instruments—trumpet, violin, viola—from large, bass instruments—bassoon, French horn, trombone—rather well. Even though the filter and the source vary together in these cases, the small-to-large correlation with Plomp's first dimension lends some support to the SMALLNESS dimension.

In a study of vowel-like sounds, Klein, Plomp, and Pols (1970) showed that listeners' confusions of these vowels, like the dissimilarity judgments of musical instrument sounds of Plomp (1976), could be represented in a three-dimensional space.[10] In this case it is possible to compare the obtained perceptual dimensions to the theoretical dimensions of sound color proposed in Rule 2. The first dimension of Klein, Plomp, and Pols appears to follow the ACUTENESS dimension closely,

---

[10] The stimuli were 100-millisecond excerpts from vowels spoken by Dutch speakers in a consonant context: [h (vowel) t].

and (with the exception of the [aw] sound) the second perceptual dimension coincides approximately with OPENNESS.

Using multidimensional scaling techniques, Shepard (1974) too has studied vowel confusions, in data from the classic experiment on vowel acoustics by Peterson and Barney (1952). Shepard also found three dimensions, the first two of which closely correspond to the ACUTENESS and OPENNESS dimensions. Shepard's third dimension seems to distinguish LAX from non-LAX vowels; however, the highly non-LAX [uu] is placed very high on the third dimension, along with the LAX vowels.

Bismarck (1974a) used the semantic-differential technique to study thirty-five synthesized, complex sounds. Listeners rated each of the sounds on each of thirty scales of verbal opposites (e.g., rough-smooth, pure-mixed, compact-scattered) and a factor analysis was performed that identified four factors or dimensions. These factors accounted for nearly 90 percent of the variance in the judgments of the listeners. Only the first of the factors was characterized by a fair unanimity of individual ratings. Bismarck called this factor *sharpness*. He interpreted his findings as suggesting that sharpness accounts for the majority of the attributes of timbre. Both musicians and non-musicians were used as listeners, and although there was general agreement among these two groups, there were significant differences as well. The musicians' results suggested that raising the high frequency cut-off of the spectrum envelope only sharpened sounds, whereas the results from non-musicians suggested both a sharpening and a scattering accompanying that change.

Only five of the thirty-five sounds studied by Bismarck were vowel-like, and these were only moderately spread on the two-dimensional plot—dull-sharp versus compact-scattered—of the results obtained from musicians (see Figure 31). This result does not argue against the importance of the F-pattern in characterizing sound color. On the contrary, Bismarck's stimuli included sounds with very abrupt edges at the cut-off frequencies of their spectrum envelopes, often at high frequencies compared to, say, the second formant of the vowels he used. Sounds whose spectrum envelopes rise in intensity to that abrupt cut-off were judged sharpest in general. One would expect that

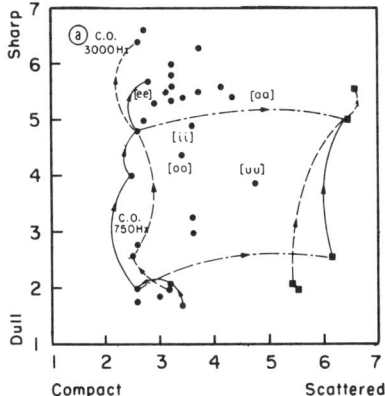

FIGURE 31: *Two principal factors of the perceptual space relating a variety of stimuli. The vowel-like sounds are identified, as are the stimuli with spectrum envelopes rising to cutoff points of different frequencies. These results were obtained using musically trained listeners. (From Bismarck 1974a.)*

sounds with a first resonance at such high frequencies—such as would be produced in cavities two inches long or less—would be judged to be as sharp as the artificial stimuli Bismarck used. If Bismarck had included among his stimuli F-patterns typical of the vowels of children, he might have been able to test whether vowel-like sounds might be more widely distributed in the two-dimensional representation than they were in the actual results.

Though the sound-color dimensions postulated by Rule 2 are not equivalent to any of the factors identified by Bismarck, his dull-sharp dimension appears to be a kind of average of the ACUTENESS and SMALLNESS dimensions of sound color. The musicians judged the highly ACUTE [ii] to be sharper than the [uu], but the [ee] and [aa] as slightly sharper than the [ii]. The non-musicians judged all the vowel-like sounds to be about equally sharp. The second factor, that of compact vs. scattered, appears primarily to distinguish periodic from noise excitation—a characteristic that, according to the present study, belongs not to sound color but to the source. Bismarck's inclusion of a wide variety of highly artificial sounds without clear resonance patterns made it unlikely that his experimental technique would reveal dimensions of sound color that are defined in terms of F-patterns.

Nevertheless, his study provides some support for certain of the sound-color dimensions.

A multidimensional scaling study by Miller and Carterette (1975) tested listeners' ability to distinguish separate dimensions in judgments of the similarity of pairs of sounds. In the first of two experiments, the results suggested that differences in fundamental frequency were more important determinants of similarity than differences in spectrum. Two dimensions related to the temporal envelopes of the sounds were also identified. In a second experiment, Miller and Carterette eliminated differences in fundamental frequency and (unfortunately, from the point of view of the present study) spectrum envelope in order to prevent the highly salient pitch differences from obscuring the effects of other variables. This second experiment was primarily concerned with the effects temporal envelopes and the number of harmonics have on the similarities of sounds.

The relevance of the Miller and Carterette experiment to the present study is largely cautionary. As those authors point out, multidimensional scaling techniques are highly unstable in the presence of a particularly salient dimension. Only by investigating acoustic dimensions of approximately equal saliency (as determined by pilot studies) can listeners be induced to attend to subtle aspects of the sound in their similarity judgments.

Dimensions of Musical-Instrument Timbre

In the experiments dealing with the dimensionality of timbre that have just been discussed, the sounds were more or less unnatural, either in their spectrum envelope or in their temporal envelopes. A series of experiments by Grey (1977) and Grey and Moorer (1977) were performed with "resynthesized" sounds made from the results of detailed analyses of the sounds of musical instruments. The stimuli in these experiments were very similar to the sounds of the musical instruments from which they were derived. Among the methods used were multidimensional scaling techniques applied to judgments of similarity among pairs of sounds. The studies agreed in identifying three dimensions of instrumental timbre. Two of the dimensions were related to temporal aspects of the sounds. The third dimension was spectrally related, with sounds having "narrow spectral bandwidth and a concentration of low-frequency energy" (French horn and muted violoncello) at one extreme of this dimension and sounds with "a very wide spectral bandwidth and less of a concentration of energy in the lowest harmonics" (oboes and trombone) at the other extreme (Grey 1977). This dimension resembles, to some extent, the sound-color dimension of ACUTENESS.

Grey (1978) extended his research by studying resynthesized musical-instrument sounds in a musical context. He measured the discriminability of sounds derived from the clarinet, the bassoon, and the trumpet, under three different conditions: in isolation, in a single-voiced melody, and in a contrapuntal setting. He concluded that a musical context tends to emphasize spectral differences among instrumental sounds, but that temporal differences are more apparent in isolated sounds. This finding suggests that the well-known dependence of instrumental timbres on attack transients (Saldanha and Corso 1964) may be weakened in a musical context.

A recent review paper by Risset and Wessel (1982) presents these and other empirical studies of timbre in musical instruments and argues for the construction of timbral spaces based on an "analysis-synthesis" approach using computer-aided synthesis techniques.

Since the fundamental frequency in most musical instruments is varied by changing the filter, instrumental sound color varies with pitch. This makes it understandable that studies of the sounds of musical instruments fail to give clear evidence for the dimensions of sound color. On the other hand, synthetic sounds produced with arbitrary, "straight-line" spectrum envelopes of various shapes result in sets of disparate, quite unnatural stimuli. Again, it is not surprising to find only a hint of evidence for the postulated color dimensions in the results of experiments using such sounds.

The problem seems largely to result from taking too empirical an approach. Without a prediction about the dimensionality of the phenomenon under study, there is little basis for the choice of stimuli. The theory of sound color suggests a program of studies using a variety of methods—perhaps including multidimensional scaling and semantic-differential techniques—on sounds having resonance peaks that are varied systematically in frequency. Interestingly, such sounds may well be perceivable as instrument-like, for a number of musical instruments have spectra with fixed-frequency peaks (Fransson 1966; Jansson 1966; Sundberg and Jansson 1976). Until such a research program is well under way, the evidence pertaining to the dimensions of sound color will remain equivocal even though the results using vowel-like sounds (e.g., Klein, Plomp, and Pols 1970; Bismarck 1974a) are somewhat encouraging.

PSYCHOACOUSTICS AND THE SOUND-COLOR OPERATIONS

No direct evidence pertaining to the operations on sound colors postulated in Rules 3a and 3b is provided in the psychoacoustic literature. However, at least one technique, worked out by Rumelhart and Abrahamson (1973) and applied to the analysis of timbre by Ehresman and Wessel (1978),[11] may cast light on psychologically "natural" sound-color operations. Ehresman and Wessel studied "timbre analogies" in a two-dimensional perceptual space derived from studies having dimensions called "spectral variation" and "brightness." Listeners were presented with a pair of sounds, A and B, a third sound, C, and then a set of four sounds, D1, D2, D3, and D4. Their task was to judge timbral analogies by ranking the four D's in terms of how well they fit the formula "A is to B as C is to D." Among the four choices for D was a sound close to the ideal solution to the analogy based on a parallelogram in the perceptual space and three others that departed from that ideal. The outcome of the experiment slightly favored the parallelogram hypothesis over a series of alternatives. Since the stimuli of Ehresman and Wessel were chosen with reference to their parallelogram hypothesis, alternative hypotheses cannot be conclusively ruled out. Also, on the grounds that the resynthesized musical-instrument sounds used in the experiment were not evenly distributed in the space and the goodness of fit of the space to the similarity judgments on which it was based was not particularly high, the authors themselves

---

[11] This study is reviewed in Wessel (1979).

suggested that their experiment should be repeated using more stimuli and more dimensions in the timbral space. The study might be improved if, in addition to the ideal sound corresponding exactly to completion of the parallelogram, sounds satisfying other hypotheses about how the analogies would be constructed were included among the choices.

Ehresman and Wessel's study suggests ways of testing certain of the operations on color proposed in Chapter Two. The perceptual salience of transposition, for example, might be demonstrable using similar techniques. The B sound color, in such an experiment, could be the A sound color transposed with respect to ACUTENESS by an amount x. The C color would be followed by a range of D colors that include an ACUTENESS transposition of C by the degree x. Some support for the dimensions themselves would be adduced if it were found, for example, that analogies in which differences between the stimuli were parallel to a single dimension—"rectangles," that is—were easier to judge than those in which differences between the A's and B's involved two or more of the theoretical dimensions.

The operations on sound color, whether or not they are found to be in any sense natural, are likely to involve processes that are commonly thought to be cognitive rather than sensory or perceptual. These aspects of the sound-color operations will be discussed in Chapter Five.

CONCLUSIONS AND OPEN QUESTIONS

Psychoacoustic studies have dealt with color, or timbre, to a greater degree than have physiological studies. There is considerable evidence that sound color and pitch should be treated as separate, independent aspects of sound. That evidence includes the following: measurements of sensitivity to frequency changes in the fundamental frequencies of periodic complex sounds is about an order of magnitude greater than sensitivity to resonance-frequency changes; and direct comparisons of the effects of different fundamental frequencies and different spectrum envelopes in sounds synthesized according to the source-filter model have shown that these two aspects of sound are mutually independent.

Psychological studies of the dimensions of timbre provide some evidence for certain dimensions of sound color. It seems likely that multidimensional scaling and semantic-differential techniques may prove adaptable to detailed studies of the dimensions. A recent experiment with timbral analogies suggests a way of studying sound color operations, although no empirical evidence for the operations can be cited.

A number of problems that seem amenable to psychoacoustic techniques remain to be solved—for example:

Sensitivity in the "stretched" sound-color space. Much research is called for to fill out our knowledge of discriminability of changes in the frequencies of resonances. One important issue is determination of the differential thresholds for resonance frequencies outside the normal range of vowels. An experiment in this area might indicate the discriminability of resonance changes in the "stretched" F_1–F_2 plane of sound colors that require normalization.

As the fundamental frequency of periodic sources rises, the frequency sampling of the spectrum envelope becomes increasingly sparse. The point at which sensitivity to changes in the spectrum envelope begins to break down as the *source* frequency increases—both in the "normal" space and in spaces having various degrees of "stretching"—should be determined as a matter of practical importance for composers.

Further psychoacoustic study of the dimensions. The existence of the postulated dimensions of sound color remains to be verified. Several techniques are available, but they will have to be applied with sensitivity and perhaps in unusual ways to reveal these subtle aspects of sound color.

Deriving scales of sound color. The development of psychoacoustic scales of sound color is of both theoretical and practical significance. As a first priority, an attempt should be made to relate psychological distance along each of the dimensions of sound color to changes in the F-patterns of stimuli. Then the form of the psychological scale itself should be investigated. We can guess that the dimensions of ACUTENESS, OPENNESS, and SMALLNESS are what S. S. Stevens (1951) has called "metathetic" continua containing only an arbitrary zero point. The LAXNESS dimension may be more like the intensive dimensions of loudness and visual brightness—"prothetic," in S. S. Stevens's terminology—since it has a natural zero point.

Studying the sound-color operations. Among the most challenging problems presented to psychoacousticians by the theory of sound color is the investigation of the transformations of sound color postulated in Rules 3a and 3b. Because those transformations represent complex kinds of auditory invariances to the listener, the task of studying them scientifically is a particularly subtle one. Their psychological naturalness is an open, and most interesting, question.

CHAPTER FIVE

Evidence from Speech and Cognitive Science

AUDITORY PHYSIOLOGISTS and psychoacousticians typically stress study of the auditory system in isolation. They ask questions about how the system is organized and how its parts and the system as a whole function, but they seldom ask how the system serves the organism in its daily activities. However, another group of scientists is very concerned with how the auditory system is put to use. Typically, in this latter group, efforts are focused on the forms of human and animal behavior that involve hearing. The sensory systems interest these researchers only to the extent that those systems serve the particular behavior pattern that they are studying. They use a broad range of methods and procedures, including those of psychoacoustics and physiology, but the aim is to understand the activity, and only secondarily the sensory or motor system that serves that activity.

One of these activities, human speech, has been studied intensively by researchers from several different fields. Not surprisingly, many of the results of speech science—a convenient generic term for all scientific studies of speech—are particularly relevant to the subject of this book. In fact, because the Rules presented in Chapter Two are derived, in large measure, from acoustic phonetics and linguistic theory, speech science almost automatically provides considerable support for the theory of sound color.

Caution is called for, however, in applying the results of speech science to the perception of sounds that are not speech-like, or of speech-like sounds that are not embedded in a speech context. If a perceptual regularity is identified in an experimental setting that encourages the listener to recognize the stimulus as a speech sound, that regularity may not be generalizable to a musical context. Certain studies of animal perception, as well as a particular line of research by speech scientists themselves, have helped to define the conditions under which these context effects may occur.

LIMITATIONS: "ACTIVE" THEORIES OF SPEECH RECOGNITION

The most influential theories of speech perception developed since the 1950s hypothesize that in order to understand speech, listeners must adopt, more or less automatically, a special attitude or "set" called the *speech mode*. Those theories—the motor theory, proposed by researchers at Haskins Laboratory (Liberman et al. 1967) and the analysis-by-synthesis theory, proposed by K. N. Stevens at MIT (1960, 1972)—are called active because they assume that in some sense speech *production* is involved in the process of perceiving speech.

The motor theory suggests that the listener to speech "talks" covertly, mimicking the talker to whom he is listening. Since the listener knows what he is "saying," he knows what the talker has said. The analysis-by-synthesis theory differs from the motor theory largely in its detailed specification of how the listener's inner "talking" leads to speech recognition. When in the speech mode, according to the analysis-by-synthesis theory, the listeners generate from memory a linguistically plausible utterance as a first guess about what is being said by a talker and then match that guess against what their auditory systems have actually received. If the difference between the guess and the auditory input (called the error score) is small enough, listeners assume that the utterance they have generated internally is what the talker has said. If the error score is above the threshold of acceptability, listeners make a second guess in an attempt to reduce the error, a second error score is calculated, and the process repeats. The analysis-by-synthesis theory does not postulate an actual motor process, but it appeals to an internal representation of language—the same one that drives the speech-production mechanisms—in order to synthesize the guesses.

In both these theories, two capabilities on the part of the listener are assumed. Listeners must first be capable of generating the internal representation; they must be able themselves to use the language of the talker.[1] Moreover, they must be able to determine whether or not a sound they are hearing is speech so that they will know when to begin the "motor" or "synthesis" process. The proponents of these "active" theories of speech perception argue that biological mechanisms specific to *homo sapiens* are required to support those two capabilities.

The two theories can be recognized as versions of modern theories of pattern recognition in general (cf. Lindsey and Norman 1977, chapter 7), although they are distinctive in that they assume a common memory structure underlying both speech production and speech perception. Sanders (1977), in a critical review of these and other theories of speech recognition, suggests that "feature matching" processes (Fant 1967) should be made a part of the models. At the present time, however, most researchers in speech perception recognize that listeners to speech cannot be passive categorizers of the speech signals. The remarkable ability of human beings to recognize utterances in a languge they know—even when spoken by strangers and sometimes in very noisy environments—requires some kind of active process on the part of the listener.

Evidence Concerning the Speech Mode

For the purposes of the present study, we are not concerned directly with the speech mode and the active theories per se but, rather, with the pre-linguistic or auditory level of sound processing. In other words, we want to know what kinds of perceptual regularities are processed at a level of the auditory system prior to the invocation of the speech mode. If we grant that something like a speech mode exists,

---

[1] They need not be able literally to talk, however. Lenneberg (1962) has demonstrated speech understanding in a patient who had never developed the ability to talk.

then we need to define its limits in order to determine what portion of the research on speech perception can be applied to questions about sound color. A great deal of research has been carried out, first, to demonstrate that the speech mode exists, and second, to define the conditions under which it is activated.

Categorical Perception

Researchers at Haskins Laboratory have presented, in support of the existence of a speech mode, evidence for what they call *categorical perception* (Liberman et al. 1967). An important part of this evidence comes from psychoacoustic measures of sensitivity to changes in certain parameters of the speech signal. It was found that people attempting to recognize a range of acoustically different, synthetic speech sounds exhibit increased sensitivity to the differences among the sounds that are near the boundaries between phonemic categories. In other words, they are better at hearing acoustic changes that make a difference linguistically than those that do not.

The Haskins experiments involved synthesized speech-like sounds having changing F-patterns, called *formant transitions*. The direction and duration of these transitions are critical cues to the identities of many consonants, including the "stops." Typically, a number of different formant transitions that sampled the range between two stops, say /b-/ and /d-/, were studied. Under ordinary circumstances, one would expect listeners to be about equally sensitive to the differences in formant transitions among these stimuli. However, the results obtained from these experiments show that sensitivity rises dramatically when the formant transitions are close to the midpoint between /b/ and /d/—where, in the speech mode, a categorical distinction must be made. These and other results have been interpreted as evidence that phonemic distinctions can be understood more clearly on the basis of articulatory rather than acoustical differences (Liberman and Pisoni 1977). In other words, motor processes (movements of the tongue, etc.) can be said to be involved in perceiving speech sounds.

Harlan Lane (1965) criticized the Haskins group's interpretation of the categorical-perception experiments by demonstrating similar effects with a series of studies in which subjects practiced intensely making arbitrary distinctions among non-speech sounds. Lane argued that the rise in sensitivity he found at the arbitrary categorical boundaries meant that categorical perception is not peculiar to speech perception but can be learned at any point in an acoustic continuum. Some more recent experiments with non-speech sounds that vary in the abruptness of their attack transients indicate that listeners classify these sounds categorically in a manner similar to that found by the Haskins group in consonant perception (Cutting and Rosner 1974; Cutting, Rosner, and Foard 1976). These results appear to weaken the case that consonant perception is handled by a specialized biological mechanism.

However, in a rebuttal to Lane's paper, the Haskins researchers (Studdert-Kennedy et al. 1970) pointed out that the category effects in Lane's results were not always as large as those found with speech-like stimuli, and suggested that his results were incomplete demonstrations of categorical perception. In that paper the Haskins group emphasized that their formulation of the motor theory applied only to "highly encoded" consonants rather than to the "less complexly encoded" vowels. Cited as evidence were the less pronounced changes in sensitivity at category boundaries in vowels and in the non-speech sounds studied by Lane, compared to the sensitivity changes in sounds with consonant-like formant transitions. This evidence is consistent with the widely held view that the speech mode is characterized by a tendency to perceive the sound stream as broken up into units (Lehiste 1972), whereas non-speech modes are said to involve continuous perception (Sanders 1977, 122). Arguments about the motor theory and the speech mode notwithstanding, there appears to be general agreement that vowels are analyzed with considerably less involvement of motor or speech-mode processes than are consonants.

Hemispheric Specialization: Clinical Results

There has long been an interest in asymmetries in bodily functions—"handedness" is the most obvious example—that appear strikingly in man and have been found to some extent to other animals. Evidence, largely clinical in nature, has shown that these functional differences between the left and right sides of the body in man are reflected in marked differences between the right and left sides of the brain. Speech has long been known to be "lateralized" to the left hemisphere of the brain.[2] For example, injury to the left temporal lobe in adults usually leads to severe and irreversible language deficits of various kinds, whereas comparable injury to the right hemisphere causes little or no such permanent aphasia (review in Lenneberg 1967).

Hemispheric Specialization: Dichotic Listening

Experiments with competing sounds fed to the two ears—an experimental protocol called *dichotic* listening—have been interpreted as support for the view that language is lateralized to the left hemisphere. These experiments show that speech sounds fed to the right ear are heard statistically more often than competing speech sounds fed at the same time to the other ear. This result can be interpreted as reflecting the brain asymmetry for speech, because (as indicated in Chapter Three) the neural pathways from the ears are mainly projected contralaterally, the right ear to the left side of the brain and the left ear to the right side of the brain. The effect is strongest with initial consonants, somewhat weaker with consonants at the end of syllables, and not statistically significant in vowels (Studdert-Kennedy and Shankweiler 1970).

The distinction between consonants and vowels in the results of Studdert-Kennedy and Shankweiler (1970) was interpreted by them to be important corroborative evidence for brain laterality for speech, because the two types of speech sounds have different linguistic and acoustic properties. The auditory parameters of speech—formant frequencies in steady-state vowels, for example—are analyzable by

[2] A more accurate statement would be that the lateralization is to the "dominant" hemisphere. In some individuals the dominant hemisphere is the right, and language is lateralized to that hemisphere. When in this chapter something is said to be a "left hemisphere" function, the dominant hemisphere is what is meant. In experimental studies that usually is in fact the left hemisphere, because potential subjects who exhibit signs of right hemispheric dominance or mixed dominance typically are eliminated from the experimental cohort.

either hemisphere, whereas the linguistic features of the signal—those associated with consonants—can be extracted only by the hemisphere that is language-dominant. The vowels, because they tend to be acoustically continuous, do not invoke this linguistic analyzer. Evidence that liquids (l, r, w, j, etc.), which are sometimes called semivowels, exhibit a small right-ear advantage (Haggard 1969) suggests that even relatively slow formant movements may invoke the linguistic mode. The characteristics of the source seem to be irrelevant: a study of noise-excited consonants (i.e., fricatives) found no right-ear advantage until a formant transition was included in the sound (Darwin 1969).

Dichotically presented musical sounds were originally studied as controls for the studies of dichotic speech sounds, then as presenting interesting questions in their own right. Dichotic melodies were found to exhibit a left-ear and, therefore, a right-hemisphere advantage (Kimura 1964, 1967). Gordon (1970) showed that competing single chords were more often identified in the left ear as well, but failed to replicate Kimura's finding of a left-ear advantage for melodies. Gordon attributed the discrepancy to the lack of contrast in timbre and loudness between the dichotic stimuli in his experiments. A left-ear advantage was found for sonar signals and environmental sounds (Curry 1967). In a melody-recognition task, Bever and Chiarello (1974) found the expected left-ear advantage in non-musicians but a right-ear advantage with practiced musicians. Papcun et al. (1974) found a shift in ear superiority for Morse-code signals in naive listeners as dichotically presented sequences of signals increased in length. Fewer than seven elements resulted in a right-ear advantage; greater than seven, a left-ear advantage. Unlike experienced Morse-code operators, who retained a right-ear advantage, the naive listeners were not able to analyze signals that exceeded seven elements in length. It was reasoned that because the long sequences could not be processed linguistically, the naive listeners had to resort to gestalt-like, holistic, right-hemisphere-mediated processes.

The weight of the evidence from experiments with dichotic stimuli supports the view that language processing—in particular, the perception of consonants in speech—requires specialized mechanisms that appear to be localized in the dominant hemisphere of the cerebral cortex. Discrimination of non-speech stimuli, including some musical sounds, may require different, non-dominant-hemisphere mechanisms, but the weight of evidence is against the idea that there is one "language and speech" hemisphere and another hemisphere for everything else. Gordon (1982) has attempted to clarify the status of the evidence for the lateralization of music in a recent review, but the thorny issue of hemispheric specialization is far from settled (Bradshaw and Nettleton 1981; Lauter 1983).

One conclusion is strongly supported by the results, however: vowels—at least steady-state vowels—are exempt from laterality in either hemisphere. The perception of vowels and, by implication, sound color seems to be less complex than the perception of consonants. Vowels are apparently discriminated at brain centers below the cortex, where asymmetries in function are nonexistent or less prominent. The evidence reviewed in Chapter Three indicating that a representation of F-patterns can be found in the auditory nerve (Sachs and Young 1980) and the indication that vowel feature detectors may be present in the thalamus (Keidel 1974) are both in agreement with the conclusion that vowels are processed below the laterally specialized cerebral cortex.

Perception of Speech in Animals

One way of investigating the limits of the speech mode in the perception of speech and speech-like sounds is to study the capabilities of other species of animals to discriminate such sounds. If an animal can be shown to perceive certain features of speech sounds, then those features, it can be argued, belong to the category of auditory rather than linguistic properties.

J. D. Miller and his colleagues have trained chinchillas to discriminate among certain human speech sounds. The chinchilla can discriminate between [i] and [a] in the face of differences in speaker, pitch level, pitch contour, and sound-pressure level (Burdick and Miller 1975); they can distinguish between syllables differing in the initial consonants [b], [d], and [g] (Miller and Kuhl 1976); and they "respond as though an abrupt qualitative change occurs . . . [at] precisely the place where many languages separate two phonemic categories (voiced vs. voiceless)" (Miller 1977). This kind of categorical perception has been demonstrated recently in the chinchilla for a distinction between voiced stops ([b, d, g]) and voiceless stops ([p, t, k]) (Kuhl 1981). Dogs can also discriminate between [i] and [a] vowels even when trained with one set of fundamental frequencies and tested with another (Baru 1975).³ It has been shown (Hienz, Sachs, and Sinnott 1981) that pigeons and blackbirds can discriminate among the vowels [a], [ɔ], [e], and [ae].

In summary, studies of categorical perception, of hemispheric specialization in the cerebral cortex, and of the perception of speech sounds in animals provide considerable, but rather ambiguous, evidence of a special mode involved in aspects of speech perception. All these studies clearly indicate, however, that vowels—and possibly a number of more complex, speech-like features of sound—are *not* perceived by means of such a special mode. The evidence is consistent with the view expressed in Chapter Two that vowels are particular cases of a broad class of natural sounds that are produced by a source exciting a filter. Apparently man shares with many other species of animals the capability of using the structural information about the environment that the present study claims is encoded in the color of a sound.

PERCEPTION OF VOWELS AND SOUND COLORS

Adoption of speech as a model for the theory of sound color does not deny that the human uses of language require innate, species-specific structures. Rather, it assumes that the perception of speech is based, at a peripheral level, on an analyzing mechanism that is phylogenetically prior to the development of language-production mechanisms. The evidence generated in response to questions about the speech mode suggests that processing in what has been called the auditory

---

³This research has been reviewed by Miller (1977) and by Marler (1977).

mode is powerful and elaborate. The vowels and sound color (as defined in Chapter Two) are both phenomena of the auditory, pre-speech mode. It follows then that we are justified, in general, in applying the results of studies of vowels to sound color.[4]

Vowel Perception in Man

The invention of the sound spectrograph during the 1940s made possible many detailed studies of the acoustics of speech sounds. Among the best known of these is a study of the vowels of American English carried out by Peterson and Barney (1952) at the Bell Telephone Laboratories. Utterances of the words *who'd, hawed, hod, had, head, hid, heed, hood*, and *heard*, by a group of American men, women, and children, were recorded and judged by a panel of listeners. The vowel portions of the words that were correctly identified by all members of the listening panel were analyzed with the sound spectrograph. The results were presented in a number of ways, the most compelling of which was a graph with the frequency of the first formant on the horizontal axis and the frequency of the second formant on the vertical axis. This "vowel diagram," with its ovals enclosing the F_1–F_2 loci of the various vowels, reveals the primary determinants of vowel color (see Figure 32).

Fant's study (1959) of Swedish vowels resulted in similar patterns of F_1 and F_2 for analogous vowels. Contrasts between the Swedish and American English vowel system account for most of the differences in the results: the /u/ vowel in Swedish is very strongly rounded and is thus lower in both F_1 and F_2 than the American /u/; Swedish has

---

[4]One exception to this rule should be mentioned. In normal human speech, steady states seldom, if ever, occur. Vowels are surrounded by consonants, and the F-patterns of vowels are almost always changing. Verbrugge et al. (1976) have presented results from experiments on vowels in consonant contexts which suggest that the speed and size of the formant transitions of the surrounding consonants affect the identity of the vowel. Research using "dynamic" vowels may have to be applied cautiously to questions about sound color. The study by Miller and Sachs (1983), discussed in Chapter Three, indicates, at least, that such formant transitions are encoded at the level of the auditory nerve.

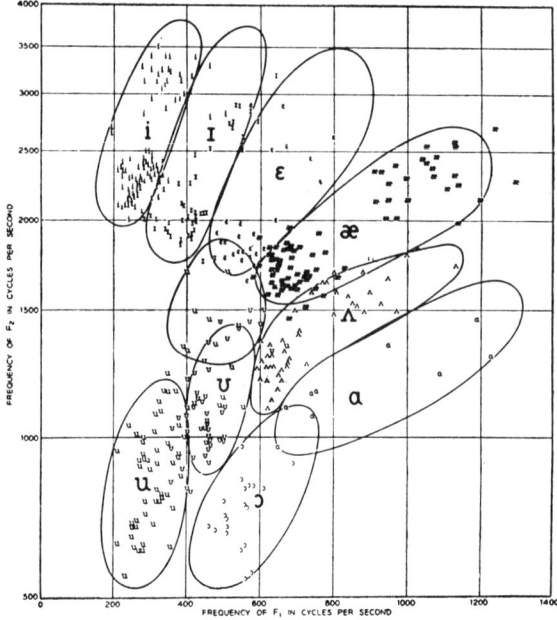

FIGURE 32: *Vowel diagram with ovals enclosing the vowel areas. F_1 = abscissa; F_2 = ordinate. The data points represent the measured formant frequencies of the vowels that were correctly identified by a panel of listeners. (From Peterson and Barney 1952.)*

rounded vowels between /i/ and /u/ that do not occur in English, and the vowel in *heard* has no close Swedish equivalent. Fant has provided a direct comparison of his results with those of Peterson and Barney (see Figure 33). It can be seen in this comparison that the third formant changes only slightly from one vowel to another in both languages. The exceptions are in /i/, where F_3 rises to about 3.0 kHz in both languages, and the English vowel in *heard*, which has an unusually low F_3.

These measurements of the F-patterns of vowels help to make Rule 1 of the theory of sound color specific. First, the measurements form the basis for the use of vowels in talking about sound color. We can imagine a sound color by imagining a vowel. The vowels serve for sound colors the same function that note names or solfège syllables serve for pitches. Second, the data of Peterson and Barney and of Fant present a greatest lower bound on the size of the sound-color space. In the vowels of the children studied by Peterson and Barney, first for-

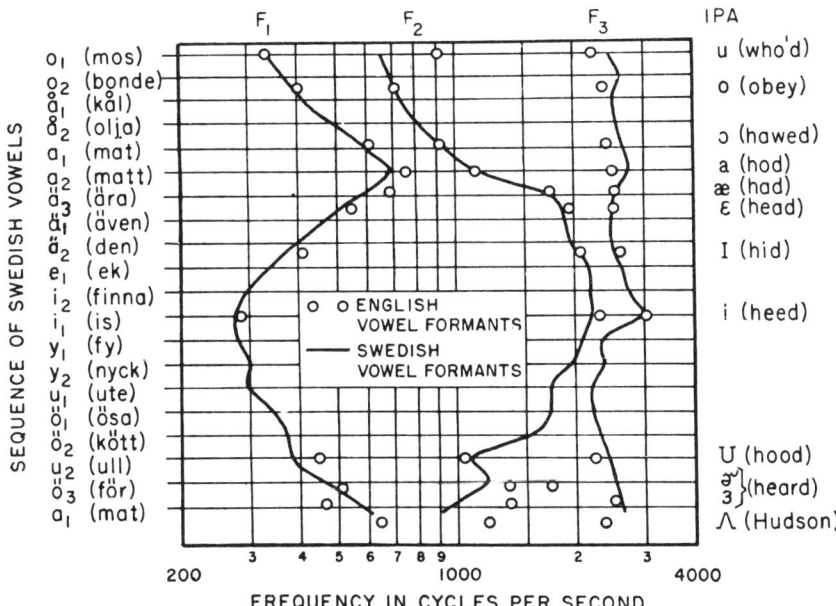

FIGURE 33: *Comparison of the F-patterns of American English and Swedish vowels. (From Fant 1959.)*

mants as high as 1.3 kHz were found (/a/ in *hod*) and second formants as high as 3.7 kHz (/i/ in *heed*). Since the formants in those vowels are a factor of about 1.75 greater than for the average adult male versions of the same vowels, a great deal of normalization was required to process them correctly. There is little doubt that some kind of normalization actually took place, because the panel identified all the vowels included in the study unanimously. The sound-color space may exceed the vowel-color space,[5] but in any case the sound-color space can be assumed to include F-patterns that are as high in frequency as the children's vowels in the data of Peterson and Barney.

Normalization of Vowels

One striking feature of the Peterson and Barney data is the overlap between certain of the ovals on the F_1–F_2 plane. In other words, the

---

[5] Informal exploration of extreme F-patterns by the author suggests that shifts of up to factors of 2 in the resonance frequencies preserve sound-color relations.

frequencies of the first two formants alone were found not to be sufficient cues to differentiate certain of the vowels. Peterson (1952) pointed out that the women's and children's vowels tended to fall in the higher frequencies within the vowel areas. He concluded that the frequency of the third formant was the most effective cue in "reducing the variance," that is, determining whether the vowel was spoken by a man, woman, or child and adjusting for that fact. The women's and children's vowels were also spoken with a higher fundamental frequency, on the average. R. L. Miller (1953) showed that the identity of certain vowels could be changed simply by raising the fundamental frequency by an octave. This result suggests that fundamental frequency may also play a role in the resolution of ambiguities in the identification of vowels. A similar small effect of fundamental frequency on vowel color was found (Slawson 1968). Using a measure of reaction time, Summerfield and Haggard (1975) studied the effects of fundamental frequency and shifted F_1 and F_3 on the identification of speakers and of vowels in words. Reasoning that pressure for quick judgments was placed on their listeners by the effort to keep reaction time at a minimum, Summerfield and Haggard concluded that the normalization determined by formant reference frames is "automatic" or, we may assume, in auditory rather than speech mode. Additional effects of fundamental frequency cues, according to these researchers, would be brought into play only at a later level, at which task-specific considerations may enter.

Temporal Aspects of Vowel Perception

Speech research gives us information about other limitations in the perception of sound color. Perception of the temporal order of a sequence of vowels can be maintained as the members of the sequence are decreased in duration down to 125 msecs per vowel; below that duration, temporal order perception breaks down abruptly (Thomas et al. 1970). If we apply this result directly to sound color, we can conclude that sequences of sound colors will retain their temporal order at speeds of as many as eight colors per second.

Vowel-like Resonances in Some Musical Instruments

As we saw in Chapter Two, most musical instruments have sources that are driven by the resonance systems of the horns, strings, or membranes that make up the instrument. There is little in those systems, apparently, that is vowel-like. However, Fransson's (1966) studies of the double-reed instruments revealed that, in addition to the tube resonances that drive the source, a broad resonance can be identified that is independent of the (coupled) reed/pipe system. He showed that this broad resonance arises in the tube that connects the mouthpiece to the body of the instrument. The center frequencies of these resonances are about 600 Hz for the bassoon, about 1.2 kHz for the oboe, and presumably about 900 Hz in the English horn.[6] These resonances may be the critical features that differentiate the sounds of the three double-reed instruments in the modern orchestra.

Jansson's studies of the acoustics of violins (1966, 1973, 1976) have also revealed a source-filter relationship that is in some ways similar to that of speech. The bow is strongly coupled to the strings, but the body of the instrument imparts peaks to the spectrum that are independent of the actual pitches played, the direction from which the sound is recorded for analysis, and the person playing the instrument. Jansson (1966) has compared the bow-string system to the vocal source and the resonance box to the vocal-tract filter system. His aim has been to describe actual violins, however, and he has not experimented with systematic variations in the characteristics of this violin "source."

Musical instruments were ruled out as models for sound color in the presentation of the theory in Chapter Two because of the strong coupling between source and filter. Now we see from the work of Fransson and Jansson that additional resonances in some musical instruments act to color the sound independently of the resonance systems that control the pitch. There may be some basis for studying the sound color of musical instruments if other decoupled resonance systems are discovered. Clearly, additional research is required on this topic.

[6]Fransson did not report actual measurements on the English horn.

DIMENSIONS OF SOUND COLOR: MULTIDIMENSIONAL SCALING OF SPEECH AND NON-SPEECH

Physiological research provides only a hint of evidence for the dimensions of sound color, and psychoacoustic research supplies only equivocal support for certain of the dimensions. However, studies of speech, suitably interpreted, provide quite strong evidence for certain of the dimensions postulated in Rule 2.

Multidimensional scaling (MDS) techniques have been applied by several researchers to judgments of vowel dissimilarity. The first two dimensions derived by Pols, Tromp, and Plomp (1973) correspond to the ACUTENESS and OPENNESS dimensions of the sound-color theory. The same two dimensions showed up in a study of confusions among vowels (Klein, Plomp, and Pols 1970) and in a factor-analytic study (Hanson 1963). The first two dimensions analyzed by Singh and Woods (1971) in steady-state vowels are also close to ACUTENESS and OPENNESS. A third dimension identified by Singh and Woods was interpreted as "retroflexion," or a curling of the tongue tip, as in the vowel in "heard." Anglin (cited in Singh and Woods 1971) found four dimensions in a study of vowels in word context. Three of these dimensions are replications of those identified by Singh and Woods; the fourth appears to be close associated with the LAXNESS dimension. Singh (1976) has suggested that the presence of the laxness feature in data derived from meaningful words, and its absense when isolated vowels were studied (e.g., Singh and Woods 1971), is due to the fact that lax vowels do not occur in isolation in English.

Presumably Singh would attribute the LAXNESS dimension of sound color to the speech mode, while exempting the other dimensions from that mode. However, the speech mode appears, on the basis of the evidence discussed at length above, to be limited to the consonants. No doubt meaningful words provide an ideal context in which to study the dimensionality of vowels; the response may be more natural than for isolated vowels. Rather than assume a specialized mode for perception of the LAXNESS dimension, we could interpret Anglin's results as suggesting that LAXNESS is simply less salient than ACUTENESS and OPENNESS and that it is hard to detect experimentally except in the ideal context of meaningful words. The retroflexion dimension corresponds to an abnormally low F_3 (see Figure 33) and may have a place as a secondary aspect of sound color. The results of Singh and Woods are a good beginning for an investigation of such additional dimensions.

These studies, in addition to those reviewed in Chapter Four, provide a reasonably strong empirical foundation for two of the four dimensions of sound color proposed in Rule 2, and weak support for the other two. The studies of vowel sounds clearly support the ACUTENESS and OPENNESS dimensions. One study supports LAXNESS. Although some studies cited in Chapter Four seem to support SMALLNESS equivocally, no research by speech scientists appears to support that dimension of sound color.

As was mentioned above, the results of MDS and factor-analysis experiments are highly sensitive to the ranges of stimuli; typically, listeners have difficulty with stimuli that vary along many dimensions. In some cases it may be possible to investigate less salient dimensions by holding the acoustic correlates of the more salient dimensions constant. When this is not possible—as in the case of LAXNESS, which generally cannot vary without changes in ACUTENESS and OPENNESS—an attempt can be made to call the listener's attention to the less prominent variation among a series of sounds that is being investigated. As indicated above, negative results in experiments using MDS or other methods do not prove that listeners are unable to perceive invariances in the attributes of sound color postulated in the present study. The results as a whole suggest that the dimensions of ACUTENESS and OPENNESS are more prominent and that composers who wish to manipulate color should be aware of that difference in prominence.

SOUND COLOR OPERATIONS:
ANALOGIES FROM COGNITIVE PSYCHOLOGY

Rules 3a and 3b, about the operations of transposition and inversion on sound colors, can be interpreted in at least two ways. The stronger of these interpretations would claim that a property of the auditory system requires that a transposed or inverted sequence *must* be heard as having an invariant relationship to the original sequence. There is little evidence to indicate that the ear functions in this way, and it would be a distortion of the intent of the theory to hold that it does. Any such automatic registration of sameness would tend to remove from the consciousness of the listener the fact that a transformation of a set of sound colors has occurred. The transposed or inverted set would be heard as identical to the original, not a variant of it, and the musical value of the operations would be reduced.

A weaker, and more plausible, interpretation of the postulated operations would hold that the result of transposing or inverting a set of sound colors *can* be heard as a version of or a variant on the original set. This weaker interpretation of the sound-color operations brings them into close coincidence with what we can imagine must be the psychological status of transposition and inversion in the pitch domain. Often a listener to a piece of music does not attend to—or misses altogether—pitch transpositions and inversions. This is particularly true of the versions of those operations that apply to atonal and twelve-tone music, in which melodic contour need not be preserved. The pitch relationships set up by application of the operations are among the aspects of a piece that are to be discovered by the attentive listener upon repeated hearings. Similarly, the operations on sound color are intended to convey subtle relationships. A kind of perceptual-cognitive process, which may be similar to that brought into play in pitch operations and perhaps in operations in other sense modalities as well, can be said to underlie detection of transposed or inverted color sets. In this kind of process, the listener is aware not only of the invariant relationship between original and transformed sets of sound colors, but also of the kind and degree of the transformation.

We can expect no direct empirical studies of either transposition or inversion of sound-color sequences, since the operations have been defined only recently, in outline (Slawson 1981). Color transposition—at least in sets that do not require wrap-around—is a straightforward analogy to transposition in pitch. Shifts along the dimensions of ACUTENESS, OPENNESS, and SMALLNESS are intuitively simple operations; they can be compared to the translation of objects in a visual field. Transposition with respect to LAXNESS is considerably more complex, but there are indications that invariances based on it can be heard. Suppose we synthesize a sequence of four colors that are widely separated in the sound-color space and then repeat the sequence over and over with gradual changes in LAXNESS. In an informal demonstration of this sort, the relationships among the four colors appear to be preserved as they are LAXNESS-transposed, and the invariance appears to survive drastic differences in the source: noise and pitched buzzes.

RECORDED
EXAMPLES
6A, 6B

Reasoning from experience with pitch operations, we might conclude that inversion of sound colors is less intuitively compelling than transposition. Nor is there any empirical evidence that demonstrates naturalness for sound-color inversion. The senses of vision and touch, however, provide models and large bodies of research on the effects of symmetrical transformations to which we can appeal. These effects, for the most part, can be shown to be independent of the particular sense modality in which they are demonstrated and may thus be cited as precedents for sound-color inversion.

Symmetry about the Vertical versus Other Transformations

Bruce and Morgan (1975) compared symmetrical and repeated random-line patterns, visually presented, in which small variations from symmetry or identity were to be detected. Variations in the symmetrical patterns were found to be easier to detect than in the repeated patterns. Kahn and Foster (1981) compared the operations of reflection and rotation in pairs of dot patterns presented at different places around the point at which the viewer's eyes were fixated. In the vast majority of the cases, the mirror image could be discriminated better than could 90° or 180° rotations. When the images were separated by a 2° visual arc, the reflected image was even discriminated somewhat better than the original image. Rock (1973) has shown that identification of symmetrical versions of a figure does not depend on a top-to-bottom axis on the retinal image but, rather, on a perception of "verticality." Mirror images were recognized as such when the viewer's head was rotated away from the vertical. Tilting objects felt by the fingers but not seen reduces their recognizability (Rock 1973, 34–36). This suggests that the detrimental effect of rotation, whether the object is seen or felt, is due to failures in assignment of direction to the perceived shape.

Each of the inversion operations on sound color represents certain kinds of symmetries. Roughly speaking, inversion with respect to OPENNESS is a reflection about the vertical axis in the F_1–F_2 plane; ACUTENESS inversion is about the horizontal axis; and SMALLNESS inversion is about the upper-left-to-lower-right diagonal. Needless to say, there is no reason to postulate a natural vertical direction in sound-color space. We have no a priori reason, therefore, to favor one sound-color inversion over another, whereas in vision the left-right mirror image is much easier to see than the up-down inversion (Rock 1973). The fact that the retinal vertical loses out to the "gravitational" vertical and that the effects are not limited to vision are grounds for supposing that analogous regularities in sound color may be identified.

Arguing against that view are the differences in detail between the neuroanatomical projections of the visual and somaesthetic systems on the one hand and the auditory system on the other (Merzenich and Kaas 1980). Whether those differences rule out symmetrical effects in hearing we do not know.

The fact that inverted melodies can be recognized is neutral with respect to this matter. Pitch inversion provides a musical precedent for sound-color inversion; however, pitch is a source characteristic whereas sound color is a filter characteristic. As has been shown at length in this and the last two chapters, these two aspects of sound behave quite differently. Actually the multidimensionality of sound color, its lack of such special relationships as octave equivalence, and the presence of a natural axis of symmetry for each dimension all make it a better analogy to the visual and somatosensory fields than pitch. Unfortunately, we have no empirical data to test the saliency of sound-color inversions. The suggestions from visual perception clearly point to a research program that might supply such data.

Mental Rotation

Starting in the early 1970s, a number of techniques have been developed to study the ability of people to rotate mental images (Shepard and Metzler 1971; Cooper and Shepard 1975; Cooper and Podgorny 1976). These studies demonstrate that subjects can rotate visual images in their "mind's eye." Reaction times were found to be linearly related to the degrees of angle through which a mental image was to be rotated. This effect is apparently independent of the complexity of the mental image (Cooper and Podgorny 1976), so the mental-rotation operation must not involve elaborate analysis of the image, which would increase the reaction time as a function of complexity. Similar effects were found for the congenitally blind, using tactile stimuli (Marmor and Zaback 1976), so the ability to rotate mental representations is not limited to the visual sense modality. These studies corroborate and, in a sense, extend and explain Rock's findings.

Sound-color inversion is not analogous to mental rotation. The inversion of a sound-color sequence with respect to one of the dimensions is a discrete mapping of colors onto colors, whereas mental rotation has a graded, analog character. Certain combinations of sound-color inversions, applied successively, produce configurations that have some similarity to those produced by rotation, but the processes are distinct. If it could be shown that sound-color inversions exhibit some of the symmetry effects of visual images, it would be worth investigating mental rotation of sound-color sequences. Such an investigation might lead to a proposal for a third operation on sound color—rotation.

CONCLUSIONS AND OPEN QUESTIONS

Although certain consonants may be perceived with reference to a specialized internal representation derived from the production mechanisms of speech, vowels appear to be discriminated by an auditory process that is more or less independent of speech. It follows that studies of the perception of vowels by speech scientists can be applied—with some caution—to questions about sound color. Those studies strongly support Rule 1 of the theory of sound color and they define a region on the F_1–F_2 plane within which the F-pattern determines sound color and is distinguishable from the source. Multidimensional scaling studies of vowels provide evidence for the sound-color dimensions of ACUTENESS, OPENNESS, and (somewhat equivocally) LAX-

NESS. There is no direct empirical support for the sound-color operations of transposition and inversion.

Clearly, a number of questions remain open. Experiments analogous to those with vowels should be carried out with sound color directly. In particular, the dimensions of sound color need study, for neither LAXNESS nor SMALLNESS is strongly supported by the available evidence. Perhaps the most provocative postulates of the theory of sound color are the operations for transforming sound-color sequences. Although here there is little direct evidence, it may be possible to apply certain techniques borrowed from studies of visual symmetries, mental rotation, and tactile perception to test whether color transposition and inversion are in any sense natural transformations.

The three disciplines that have been reviewed in Chapters Three, Four, and Five clearly would gain much by judicious cross-fertilization. Over the last decade, conscious and sustained efforts have increasingly been made to bring a unified approach to the study of the auditory system. Reviews by Symmes (1981) and Studdert-Kennedy (1979); the series of symposia on the physiology and psychophysics of hearing (Plomp and Smoorenburg 1970; Cardozo 1972; Zwicker and Terhardt 1974; Evans and Wilson 1977; Møller 1973; Syka and Aitkin 1981); similar symposia in speech research (Fant and Tatham 1975; Lindblom and Öhman 1979); and the bringing together of students of animal behavior, speech, and hearing in one of the Dahlem Conferences (Bullock 1977)—all are signs of the importance large numbers of researchers assign to an interdisciplinary approach. The scientific study of sound color cuts across all these fields. Even though at the moment there are relatively few hard facts and many questions about that aspect of sound, the growing together of the various fields—primarily by posing cross-disciplinary questions—is largely responsible for the substantial, if still woefully incomplete, empirical support that can be cited at present for the theory of sound color.

CHAPTER SIX

Musical Evidence: Sound Color in Electronic Music

THE THEORY PRESENTED in Chapter Two and the scientific evidence that supports it suggests that it is possible to treat sound color compositionally as an independent element of music. But has it been so treated? Can it be shown that composers have used sound color independently (Rule 1 of the theory) and in ways that can be interpreted as suggesting that they hear the dimensions (Rule 2)? Do they appear to transform sound color with reference to those dimensions by application of certain operations (Rules 3a and 3b)?

To answer these questions we must turn to analyses, primarily, of electronic music, because electronic music provides both the opportunity to control sound color and the possibility of choosing not to do so. The filters in electronically generated music are constrained in practice to fairly simple configurations, but within those configurations they are effectively unlimited in frequency range and they typically can be controlled with considerable ease. They are flexible devices that can be used elaborately if composers choose to do so. Sources, too, are many in electronic music. Nearly the only constraints on sources are imposed by limitations in the flexibility provided by synthesizers and computer programs for signal control. The theory of sound color can be tested better in this relatively free context of electronic and computer music than it can in vocal music, in which sound color is more or less automatically controlled according to the theory, or in instrumental music, where independent control of sound color is difficult and in some cases impossible. What do composers do when they are freed from these limitations and can control sound either in ways that are consistent with the theory of sound color or in quite different ways? If we can find evidence that the rules postulated by the theory have been followed by composers of electronic and computer music, that is strong support indeed for the theory.

In principle, vocal music could supply some evidence, particularly for the operations on sound color. The tradition of setting preexisting texts notwithstanding, there is no reason why composers could not compose vowel structures in a manner that follows from the theory of sound color. In fact at least two works, Babbitt's *Phonemena* (1977) for voice and tape or piano and Stockhausen's *Stimmung* (1971) for six amplified voices, are based in part on vowel and consonant sets. It would be meaningful to ask of these and certain other vocal works if the postulated dimensions of sound color are at all in evidence and if color transposition or inversion is employed.

LIMITATIONS OF MUSIC-ANALYTIC EVIDENCE

Each of the scientific disciplines whose evidence was cited in the last three chapters has certain limitations, as does the appeal to musical analyses. These limitations range from general problems in verification of a music theory by musical analysis to specific technical issues.

Music theorists traditionally have supported their theoretical claims by citing the music of respected composers of their own and previous generations (Glarean [1547] 1965; Morley [1597] 1953; Schenker [1935] 1979; Forte 1978). The logic of this approach depends, first, on the reader agreeing that the music cited is successful, and second, on the theory explaining something significant about the music. Theorists can do little to ensure that their readers agree with them about the quality of the music. The authors must simply select music that they themselves value and trust that the reader, to some degree at least, will share those values. The second point depends more on the explanatory power of the theory itself and the case made by the theorist. The burden of proof rests with the theorist to show that the theoretical claims hold for a number of pieces of music, and that the theory elucidates some of their important features. If the elucidation is successful and the music is well chosen, the musical relevance of the theory is strengthened. Not only is there reason to apply the theory to other music already composed, composers are encouraged to attempt to use the theory in new music.

A problem arises here that is common to all musical analyses. The discussion of musical excerpts in this chapter, like all music-analytic discourse, is not easily verifiable. The analyst describes his or her intuitions about a particular passage, but that description is valued by readers only insofar as it coincides with their own intuitions. Needless to say, general agreement by colleagues gives an analyst confidence in making analytic assertions, but they remain assertions, not facts. It is, however, precisely such analyses—descriptions of the way something sounds—that can provide the most direct, musically concrete verification of a theory like the one advanced in the present study. Since we lack a complete auditory and esthetic theory, there is no way out of the solipsism; we must simply recognize that music-analytic evidence is not scientific evidence but brings humanistic insights to bear on the theoretical claims.

A related technical issue is raised by the possibility of preparing objective graphical displays of the musical passages to be considered.[1] In most cases such graphs cannot aid the analysis of the details of sound color appreciably, because the passages in question are usually complex in texture and we have little experience in how such displays should be interpreted. Only after years of extensive experience—and then only imperfectly—can phoneticians learn to recognize arbitrary speech utterances from sound spectrographs of those utterances (Cole et al. 1980). The problem of recognizing the color of a sound in a graphic display of the total spectrum of a contrapuntally complex musical passage can hardly be appreciably less difficult. This is not to deny that such displays may prove useful as a tool for verification, but they

---

[1] Cogan (1980) has presented one form of these kinds of displays.

cannot substitute for careful analysis by ear. For large-scale analyses of the sort undertaken by Cogan, the visual displays appear to have value.

INDEPENDENT MUSICAL CONTROL OF SOUND COLOR

Rule 1 postulates how sound color can be held invariant when other aspects of a sound are changed. We can suggest that Rule 1 is followed by composers if in particular passages we hear that sound color is indeed held constant while something else about the sound is varied. Conversely, we can demonstrate the independence of sound color if it alone is heard to vary while everything else is held constant. Partly because the band-pass filter, either manually or voltage-controlled, is a common feature of the electronic music studio, a large number of electronic compositions exhibit the invariance and independence codified in Rule 1. The filter[2] imposes over the frequency spectrum of a source of sound a template that is a function, not in any way of the sound source itself, but the settings of the filter. Passages in which filtering is an important component, therefore, are examples of, and support for, the musical application of Rule 1 of the sound-color theory. Most pieces of electronic music involve filtering of one kind or another, but they vary greatly in the extent and manner of application of this correlate of sound color. Passages from some of these pieces can be classified into broad categories that illustrate certain of the ways invariance of sound color can be put to use musically.

Color as a Contrapuntal Differentiator

The first category of sound-color uses is analogous to the allocation, in instrumental music, of contrapuntal lines to different instruments. In a typical case of this musical application of sound color, each contrapuntal line is synthesized with a complex periodic waveform and then

---

[2] Many kinds of filters are used in analog electronic music studios, including manually controlled low-pass, high-pass, and bandpass filters, equalizers, and voltage-controlled filters. Filters are sometimes used in computer-synthesized music, but less often than in analog synthesis.

modified by a characteristic filter pattern that is fixed throughout a phrase, a more extended passage, or even an entire work. The different filter settings for each voice suggest that the composer's intention is to differentiate and give coherence to the separate contrapuntal lines. The sound colors in these pieces may be carefully selected, but other elements of the music, usually pitch and rhythmic patterns, clearly predominate. The changes in color are few, whereas the pitches, the intensities, and other aspects of the sound change relatively rapidly, forming the main substance of the piece.

An important early work by Milton Babbitt, *Composition for Synthesizer* (1964), is a particularly clear example of this use of sound color. Babbitt treats this first piece made with the RCA synthesizer as an extension of his instrumental music. The pitches are organized from a twelve-tone series presented straightforwardly, twenty-three seconds into the piece, entirely in one of the two channels. The "melody" here sounds as if it is played by a buzz source shaped by a fixed filter. The sound color is slightly less ACUTE and OPEN than the neutral sound color—it is a LAX [oh]. The pitch series is repeated several times at different transpositions and with successively shorter durations. Against these presentations of the melodic form of the series are other voices that contrast in apparent spatial location, register, envelope, and sound color. The consistent sound color of the successive repetitions of the "melody" is the most prominent invariance that leads one to hear these repetitions as belonging to the same strand or voice.

This is only the first of many striking uses of sound color to differentiate simultaneous contrapuntal lines in *Composition for Synthesizer*. Little in this piece suggests that sound color was manipulated in its own right as an element equal in importance to pitch and duration. The filter characteristics seem to have been chosen so as to emphasize differences among the voices, and the sound colors—at least in the melody-like strands—are typically unchanging throughout phrases and often over entire sections of the work.

Many other electronic works from the 1950s and early 1960s commonly use color as a factor in contrapuntal differentiation, but seldom as consistently as in Babbitt's *Composition for Synthesizer*. His stated interest in establishing a continuity of technique from instrumental to synthesized music (Babbitt 1962), and possibly his desire to exploit the independence of control of the generators and the filters in the RCA synthesizer, may have led him to the consistent use of this technique. The remarks attributed to Babbitt on the record jacket (1964) support this view: he expresses interest in the intricate control of pitch and rhythm made possible by the pre-programming feature of the RCA synthesizer, not in the invention of "new sounds and timbres."

The use of sound color in *Composition for Synthesizer* seems to be effective in differentiating voices, so that piece provides support for Rule 1 of the theory of sound color. The presumably fixed spectrum envelope imposed upon the buzz-like sources unifies a "melody" of changing pitches. The conditions of Rule 1 were apparently met: fundamental frequency is changed while spectrum envelope is held constant. The fact that we hear coherent strands of sound, differentiated from other strands in the texture, supports the conclusion that Babbitt is manipulating an aspect of sound—sound color—that is independent of the other parameters.

This use of sound color is very common. Almost any electronic piece that has a "note-like" character—from "switched-on" versions of instrumental music to pieces like Wuorinen's Pulitzer Prize-winning *Time's Encomium* (1969)—contain passages with filters used in this way.

Invariant Sound Color as a Structural Link

Another striking technique in which sound color is held invariant is the linking and integration of consecutive sections that contrast strongly in some non-coloristic aspect of timbre. For example, a piece may have a section in which periodic pulses dominate, followed by a section in which noise predominates. The sounds in the two sections are both passed through a filter network. The strong contrast in source characteristics is the most salient feature of the juncture between the sections, but if the filter network is kept the same across the juncture, an element of unity is provided. The constant sound color resulting from the constant filter network is a subtle link that justifies the contrast in sources. Any such structural use of invariant sound color can be construed as treating that element as an equivalence class. Establishing that composers have used sound-color equivalence across a sharp sonic contrast would strongly support Rule 1 of the theory.

Druckman's *Synapse* (1971) provides an interesting and subtle example of this use of color. Following the first section of the piece (at about 2:25),[3] a second section is formed by events in which both pitch and color change dynamically, often in "opposite" directions. The pitch of one of these events slides downward, while the color glissandos through a sequence of approximately [uu ne ae ii]. The color change is mostly in ACUTENESS or SMALLNESS, from low to high values on those dimensions. The sound was probably realized with a fairly simple control function, causing a single resonance filter to sweep from around 500 Hz to around 3.0 kHz. This section ends with a similar sound, in this case one whose pitch and color move downward together—the color sweep approximating, in order, [ii ae ne uu]. The link between this section and the next provides the example of sound color as an equivalence class.

The first sound in the next section is a medium-pitched sinusoid. Although apparently no filter was used, the sound is [uu]-like in color, probably because the lack of energy above the fundamental closely matches the spectrum of the [uu]-like color that closed the previous section. These two sections are strongly contrasting in their sources, however. The last sound of the first section is made from a source rich in partials; the first sound in the second section, from a source having only a single component. It is the filter in the first sound that strongly attenuates the higher partials and effects the link. This transition from a complex, filtered sound into a sinusoid of medium or low pitch is interesting because it is very specific: an [uu]-like sound color, which is low in ACUTENESS and OPENNESS, is required in the complex sound. In sinusoids, pitch and sound color are confounded. Depending on

[3] Time points and durations in this and the following chapter will be expressed in minutes and seconds: thus, 2:25 indicates the point in the music that occurs 2 minutes and 25 seconds after the beginning of the piece.

the perceptual "set" of the listener, either pitch or color can be heard, the two co-varying with frequency. Early studies of single formant vowels (Stumpf 1926) are of interest in this regard. In certain cases, a single sinusoid can take the place of a formant, or even a pair of closely spaced formants, in cueing sound color; Druckman has exploited this in a musically subtle way.

Stockhausen's electronic music provides many instances of color matching over a musical juncture formed by strongly contrasting sources. One that is of particular structural significance occurs in Region II of *Hymnen* (Stockhausen 1969a) at the transition into the prominent presentation of the German national anthem (with recorded orchestra and chorus; about one-third of the total duration of the movement from its beginning). The transition is begun when an OPEN, [aw]-like sound that has been heard in brief fragments over the previous minute is added intermittently to a long, low-pitched drone with an [uu]-like color. As the added sound is reattacked, its pitch is raised and its color is shifted toward the ACUTE and OPEN [ae] in preparation for the abrupt entry of the orchestra playing the familiar anthem. The "bright" sound of the orchestra and chorus is well matched by the final [ae]-like color of the electronic sound that immediately precedes it. This instance of linking by means of sound color is also of particular interest because the sound was probably produced, not with a changing filter, but with a tape-speed change. This is the relative-pitch transformation, in which pitch and color change together. The link is not effected with pitch, however. The end of the electronic sound has a rather indefinite, changing pitch that has no relationship to the definite pitches of the orchestral music that follows. It is the [ae]-like overall color of the orchestra, not its pitch, that is anticipated in this masterful connection over this important juncture. The link in this case is particularly effective—even, one could say, necessary—because the direction of Region II to this point has been toward low pitches and non-ACUTE, non-OPEN colors. The drama of the entrance of the German anthem is enhanced by the contrast with the long, low drone that precedes it. At the same time, some common ground across the juncture, desirable at any such point, is made doubly important by the strength of the contrast. The choice of sound color as the equivalence is appropriate because pitch is the most salient parameter in the anthem and even, perhaps, in the drone that precedes it. Color provides the link over the strong contrast in source characteristics.

Less common than its use as a contrapuntal differentiator, the use of sound color as a link, typically in the music of experienced composers, provides a convincing demonstration of the independence and musical relevance of sound color.

Color in the Foreground

Many electronic works employ filters actively and pervasively. This practice follows naturally from the design of voltage-controlled, analog synthesizers, in which the components are independent and, for the most part, weakly coupled to each other. The filters typically are bandpass, or low-pass, voltage-controlled filters and there is usually a single filter for each strand or contrapuntal line of the composition. The sources in these works are any of a wide variety of clicks, pulses, noises and, somewhat less often, periodic buzzes that have pitch. In this context sound color becomes a foreground element in its own right.

The music of Morton Subotnick is the best known and among the most highly developed in which sound color is in the foreground. In a series of works beginning with *Silver Apples of the Moon* (1967), Subotnick has explored many of the possibilities of filtered sound with a number of different kinds of sources. Among the most ambitious of these works is *Until Spring* (Subotnick 1976). This piece, like much of Subotnick's other electronic music, emphasizes repetitively pulsed sounds in multiple contrasting but interacting strands. Each strand of repeated pulses is nearly always filtered by a changing bandpass filter to create changing sound-color patterns. Subotnick carefully controls the sound-color ranges and directions of movement to give form both on a small scale (1 to 2 seconds' duration) and over large sections. Near the beginning of the second half of *Until Spring*, for example, the general tendency of the most prominent strand is a gradual change from [uu] or [oo] through the center of the sound-color space, ending with [ae]-like colors (see Figure 34a). With small variations and the addi-

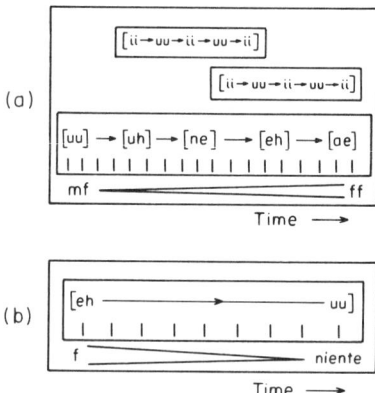

FIGURE 34: *Graphical scores of two excerpts from Subotnick's* Until Spring. *Time is the horizontal axis; the sound colors in the contrapuntal strands are represented by vowel equivalents in rectangular boxes. (a) A passage typical of the beginning of the second half of the work. At about 3:20 from the beginning of the second half of the work, a contrasting sequence of colors (b) closes the section.*

tion of secondary strands, these gestures gain intensity, speed, and complexity until, at about 3:20, a single brief phrase abruptly closes the section. This final phrase is opposite in virtually every respect to the tendencies of the immediately previous music. It is a single strand, its intensity curve is from high to low, and most striking of all, its sound-color change is opposite to what has preceded it, changing from [ae]-like or [eh]-like colors back through the neutral color to [uu]. Only the repetition rate of the pulse excitation is kept approximately the same in this final phrase (see Figure 34b).

This use of color in this piece as a primary carrier of musical information (along with the metric interactions of the various pulse rates in the contrapuntal strands) represents clear evidence of the musical relevance of Rule 1. The directions of change in sound color are similar to those in Druckman's *Synapse*, but in Subotnick's *Until Spring* the pulsed sources remain the same over long durations, and sound color carries a greater portion of the shape and substance of the music. This is an extension and corollary to Rule 1 of the theory. Once we know how to hold sound color invariant over changes in source characteristics, we can choose to hold the source invariant and change the sound color.

Once again we are dealing with a common musical application of sound color that is effective in the hands of accomplished composers. There is little need to cite additional examples. Conceptually, this rather literal kind of realization of sound-color "melody" is straightforward. The art lies in the specific choice of sound colors, sources, and temporal settings—any of which would serve equally well as evidence for the musical independence of sound color.

EVIDENCE FOR THE DIMENSIONS

Musical evidence for the postulated dimensions or attributes of sound color is—almost of necessity—equivocal. If composers conceived of the dimensions explicitly as postulated in Rule 2, their music would undoubtedly reflect that explicit conception. We would expect to find, for example, sound-color sequences varying in ACUTENESS but having only a single value of OPENNESS. Such a sequence might be followed by, or in some other way related to, a second sequence with a similar pattern of variation, but now with OPENNESS changing and ACUTENESS held constant. No clear-cut illustrations of conscious manipulation of sound color reflecting the dimensions have been found. Even if cases could be cited, we would have evidence simply that the composer of the passage in question conceived of the dimensions of Rule 2. The theoretical corroboration would be welcome, but we could claim only indirectly that the dimensions describe a *perceptual* sound-color space.

Suppose, on the other hand, that we were to find traces in a piece suggesting that the composer hears a regularity corresponding to one or another of the dimensions, but tenuously, perhaps inconsistently, indicating something less than a consciously-worked-out method. This kind of evidence would permit us to answer questions about the psychological reality of the dimensions more directly. Discovery of this tentative sort of musical application of the sound-color dimensions would permit us to confirm in a musical setting the perceptual evidence from multidimensional scaling and other kinds of scientific studies.

In the literature of electronic music there are indeed passages of this sort. The three passages cited below exploit the dimensions in different ways, and each provides a different kind of evidence.

Change along Specific Dimensions

Except in special cases, variation in sound color involves changes in the values of all the dimensions. Certain of the special cases are interesting because they approach closely the conditions of independent variability along a single dimension in the sound-color space. The main criterion cited in Chapter Two for establishing a dimension was that it could be held independently constant on equal-value contours in the space. Independent constancy permits us to admit the dimensions of SMALLNESS and LAXNESS, which are not perpendicular to ACUTENESS and OPENNESS. The latter two dimensions are (approximately) orthogonal, so one can be independently varied without changing the other. A musical passage in which changes in ACUTENESS are made while holding OPENNESS constant (or vice versa) suggests that the composer hears at least those two aspects of sound color and is manipulating them compositionally.

Subotnick's *A Sky of Cloudless Sulphur* (1980), composed in 1978, contains such a passage. Throughout this work changing, diphthong-like sound colors are strongly in the foreground as they are in *Until Spring*. The section of this work called "Cocoon" (starting at ca. 2:02) is very quiet and sparse, characterized by isolated, percussive, semi-pitched pulses associated with sounds having changing patterns of sound colors whose source of excitation is a kind of flutter of pulses. The first two of these patterns include only colors of high OPENNESS that vary in ACUTENESS from [aw] to [ae]. A third event is similar, but it begins with an [oo]-like color. Then a much longer event starts at [oo] and slowly moves through [aw] to [aa] and then dies away toward a less clear, possibly neutral color. This is followed (at ca. 2:53) by a series of color sweeps that change in the direction of increasing and then decreasing SMALLNESS (approximately [uu ne ae ne uu]). At about 3:45, the quietest spot in the piece, nearly all the colors are low in ACUTENESS. We hear first [aw] and [oo], then [uu]. Then gradually over the next minute the colors become more ACUTE as a continuous flutter enters. At 5:05, very abrupt snapping sounds are associated with [aw aa ae], highly OPEN colors that vary in ACUTENESS. Finally, at 5:54 the [ii]-like colors are reached with events that change across [aw aa ae ee ii]. The piece becomes increasingly active from this point on, moving gradually into the third large section of the work, "Butterfly."

At several points in this section there are suggestions of a preoccupation with changes in one of the two orthogonal dimensions while holding the other constant. The colors of constant high OPENNESS at the beginning of the section that vary in ACUTENESS, the colors of low ACUTENESS varying in OPENNESS in the very soft passage, and more generally, the slow opening of the sound colors over the following two minutes toward increasing ACUTENESS are all examples of this preoccupation. The last example is particularly to the point, because changes in OPENNESS are on a time scale of at most a second or two, whereas the changes in ACUTENESS range over the entire span of about two minutes.

These observations are contradicted at a few points in the section, where sequences of colors occur that include changes in both orthogonal dimensions. Moreover, an alternative interpretation could attempt to describe the sound colors in this section as based on a single dimension ranging from [uu] to [aw] to [ae] to [ii]. The occasional neutral sounds and the [uu ne ae ne uu] sequence near the beginning of the section do not fit well with this alternative, however. Although ACUTENESS and OPENNESS are not consistently treated in an independent manner in this music, there are sufficient indications in this passage—and several others throughout the work—that the composer hears these dimensions and often controls sound color in ways that reflect his perception.

Exposition of the Sound-Color Space

Babbitt's second major work realized on the RCA synthesizer in 1964, *Ensembles for Synthesizer* (Babbitt 1976), presents another kind of evidence for the dimensions of sound color. In this piece, sound color and a variety of different source types serve as contrapuntal differentiators, but to a far greater degree than in *Composition for Synthesizer*, timbre—including sound color—is an independent parameter of the music. Recognition of the dimensions of sound color in *Ensembles* is suggested in a different way than the movement along orthogonal dimensions found in Subotnick's *Sky of Cloudless Sulphur*. In exposing the palette of *Ensembles* during the introduction, Babbitt fills out the sound-color space. In so doing, he indicates something of his conception of that space and provides evidence for certain of the sound-color dimensions.

The introduction of *Ensembles* is made up of four contrapuntal passages, each ending with a more-or-less static mass of sound having a distinctive mixture of sound colors. The first of these static sounds is made up of colors generally high in ACUTENESS and low in OPENNESS. The most prominent color is [ii]-like, but the sound includes [ee]-like colors and LAX sounds close to the center of the sound-color space. The "opposite" side of the space is occupied by the second of these static sound masses, which contains colors low in ACUTENESS and high in OPENNESS. The [aw]-like color predominates, but again LAX sounds are included, as are colors corresponding to [oo] and [aa]. The primary color of the third static sound is high in both ACUTENESS and OPENNESS, with the strongest color close to [ae] and including [ee] and some LAX sounds. The upper right quadrant of the space is occupied by this third sound mass. Finally, at the end of the introduction, the last of the four sounds fills the remaining quadrant of the space: the colors around [uu] that are low in both ACUTENESS and OPENNESS, and their LAX versions. The four sounds represent a systematic filling of the sound-color space.

Spectrographic displays of excerpts from the four static sounds are presented in Figure 35. The graphs picture the four sounds in time (horizontal axis)/frequency (vertical axis)/intensity (darkness) plots. These pictures to some extent confirm the preceding aural analysis: maxima appear in the spectrum plots that correspond approximately to the expected resonance frequencies in each of the four sounds. This acoustic analysis is far from clear-cut, however. The several "extraneous" resonances indicated by the spectrographs emphasize the

RECORDED
EXAMPLE 7A

RECORDED
EXAMPLE 7B

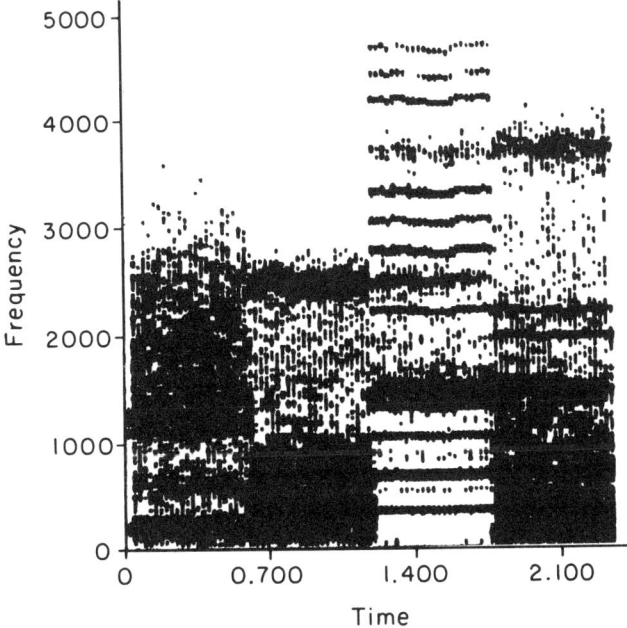

FIGURE 35: *Spectrographs of excerpts from Babbitt's* Ensembles for Synthesizer. *The four sounds are brief excerpts from the four static sound-masses of the introduction. A prominent band of frequencies from about 1.2 to 1.6 kHz appears in all the sounds, obscuring the second resonance in certain of the sounds. This constant feature may account for the observation that the sound masses include neutral sound colors.*

sounds' complexities and suggest how limited the contributions of acoustic measurements are to the detailed study of sound color.[4]

This analysis suggests both that Babbitt is manipulating sound color independently and, more importantly, that he treats color as if it

[4] As was pointed out in Chapter Two, the necessary and sufficient acoustical conditions that give rise to perceived sound colors are not completely specified by the theory. Such a specification will require an extensive *psychoacoustic* research program. Until that program is undertaken, an introspective impression of the sound colors in musical passages, with all the dangers of bias that can entail, is often better evidence than objective acoustical measurements in which the relevant acoustical parameters (the F-patterns of the most salient filter system) may be masked in the visual pictures of the sound by irrelevant noise.

occupied a two-dimensional space. The variety of sources in the four static sounds, including unfiltered sinusoids, filtered buzzes, and noise, all contribute to the colors of the sound masses, but none of the individual sources could substitute for the entire complex without altering it significantly. Certain components of the sounds are pitched, but the emphasis, it seems clear, is on the sound color of the sound-mass as a whole, not on its pitches or on the sound color of one of its components. Two aspects of the sounds pertain to the question of the dimensionality of the color space. The inclusion of LAX sounds that are perceptually similar in all four of the mixtures suggests their distribution about a central area, not along a single high-to-low line. The order in which the sounds occur also suggests the successive exposure of a space of two dimensions. We have an impression of opposites in the first and second and in the third and fourth sounds, each occupying a quadrant of the complete color space. Had the order been from [ii] to [ae] to [aw] to [uu] a better case could be made for a single dimension. As it occurs in the piece, the succession of these sounds points strongly to a conception that is consistent with Rule 2 of the theory of sound color.

The foregoing analysis by no means exhausts Babbitt's use of sound color as an organizing force in *Ensembles for Synthesizer*. The contrapuntal passages in the introduction (preceding each of the static sound masses discussed above) present particularly interesting choices of sound colors. The colors in the lines of these passages, taken as a whole, also appear to fill the sound-color space, just as the associated static sound-masses do "vertically" in the same passage. The return of a version of the introduction, interestingly modified, at the very end of the piece is a further structurally significant use of sound color. These superficial observations about the piece suggest that a detailed sound-color analysis, although beyond the scope of the present study, would be rewarding.

Completion Based on Sound Color

A musical function closely related to the systematic filling of the color space in Babbitt's *Ensembles* is use of sound color as a means for completion and closure at the end of a piece. Stockhausen's *Telemusik* is a case in point. In this work, composed in Japan during 1966 and issued in score (Stockhausen 1969b) and as a recording (Stockhausen 1970), he makes copious use of a variety of sounds from the musics of Japan, China, and Southeast Asia, which he modifies electronically and mixes with each other and with electronically produced sounds. Many of these borrowed sounds, unlike those of most instruments of the symphony orchestra, have very distinctive sound colors. As a result Stockhausen is able to integrate filtered electronic sounds in a profound and structurally significant way with the "real" instrumental and vocal sources.

The closing two phrases of the composition, which include both electronic and prerecorded sounds, apply sound color in a distinctive way to give a sense of finality and completeness to the work. The penultimate phrase contains three main strands of sound: a high-pitched, ACUTE-sounding collection of steady-state sinusoids, a vocal chorus of Buddhist chanting that centers on the sound colors [aw] and [aa], and ring-modulated Gagaku music that varies across [uu eh ae eh uu] but centers mostly on [uu]. The final phrase (presented in graphical score in Figure 36) features an increase of activity and loudness in all five of the tape-recorder tracks (each has its own horizontal rectangle in the score). Track I consists of very high-pitched, glissando sounds that are primarily [ii]-like in color, but they include [ee] and [ae]-like colors as well. Tracks II and V sound together with electronic sounds filtered to an [oo]-like color. The Balinese gamelan music in Track III is accelerated, sounding in toto with an [ae]-like color. Track IV punctuates the beginning of the phrase with a percussive stroke from a Japanese drum that evokes an [ii] or [ee]-like sound color. After eight seconds, a second stroke from a low-pitched drum punctuates the second gesture of the phrase with an [aw]-like color in Tracks II and V. Electronically produced sounds in Track IV then present another sound that glissandos from an [aa]-like color to an [ae]-like color. The final stroke in all tracks is a complex mixture of percussive sounds that suggests sound colors of [ne] and [oe]. Only in this final sound of the piece do we have a color, [oe], from the region of the sound-color space having moderate ACUTENESS and low OPENNESS. Unquestionably, the temporal place-

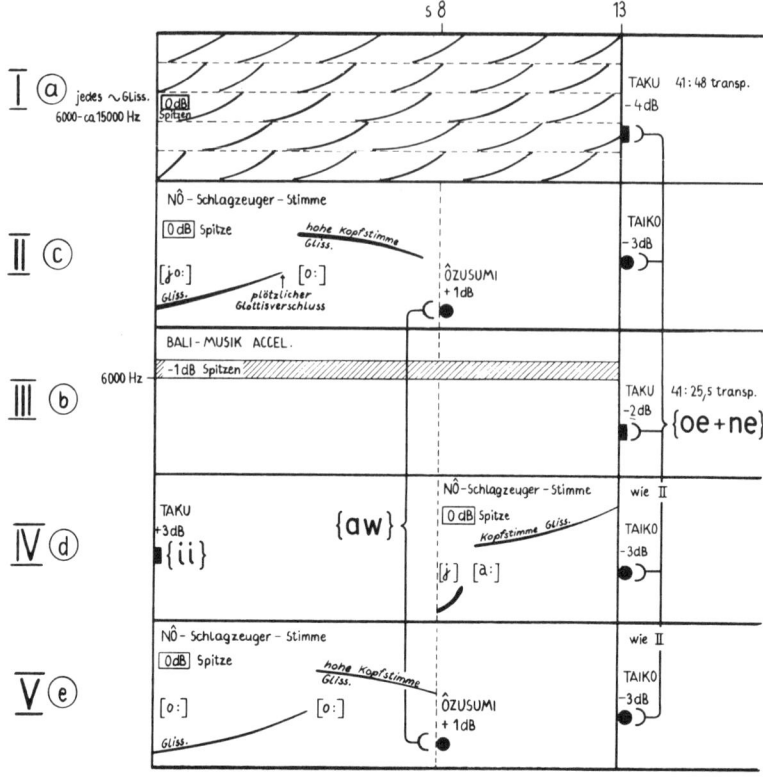

FIGURE 36: *The last phrase of Stockhausen's* Telemusik. *The sound colors in the original score are IPA symbols (enclosed in square brackets); those added to characterize the sound colors of the percussion strokes are in curly brackets. (From Stockhausen 1969b.)* © *Copyright 1969 by Universal Edition A.G. Wien,* USED BY PERMISSION.

ment of the three percussive strokes contributes strongly to the sense of completion—almost of cadence—but the color of the last sound of the composition seems also to have been chosen to complete a kind of aggregate that fills the sound-color space, reinforcing and confirming the close.

Stockhausen's use of sound color in these passages suggests that he hears a sound-color space of at least two dimensions. The penultimate phrase strongly contrasts sounds of low OPENNESS and low ([uu]) and high ([ii]) ACUTENESS with sounds of low ACUTENESS and high OPENNESS ([aw]). The high ACUTENESS and high OPENNESS area,

[ae], is hinted at with the ring-modulated sounds, which are concentrated mainly on [uu]. In the last phrase, the last glissando of the piece almost completes a color change from [aa] to [ae]. As in the excerpt from the Babbitt work, the order of occurrence of these sounds does not suggest a single low-to-high dimension. The last [oe]-like stroke is particularly significant because it does not fit easily into a one-dimensional continuum that also contains [aa]. Unlike Babbitt's *Ensembles for Synthesizer*, the electronic music of this period of Stockhausen's production does not suggest a strict serialism but, rather, an intuitive approach to composition with a set of sounds that have been selected initially at a fairly abstract level according to some systematic procedure. If we assume that Stockhausen's compositional procedure was of this kind, the passage from *Telemusik* provides evidence that may be somewhat less influenced by theoretical considerations than the *Ensembles* excerpt, even if the argument supporting the multidimensionality of sound color cannot be made quite as clearly with reference to the Stockhausen work as it can for Babbitt's.

OPERATIONS ON SOUND COLOR?

The "opposite" motion in Subotnick's *Until Spring* and at the end of the passage cited from Druckman's *Synapse* can be interpreted as instances of SMALLNESS inversion, but those sequences of sounds can also be explained by the operation of retrogression. In fact, the common use of dynamic single filters—or of simple changes in waveform that mimic such filters—produces almost automatically the [uu ne ae] sequence whose retrograde is its SMALLNESS inversion. If we consider in successive, disjunct pairs the four static sounds from Babbitt's *Ensembles for Synthesizer* that were discussed above, we can interpret the second pair, [ae] [uu], as the OPENNESS inversion of the first pair, [ii] [aw]. Although simple explanations, such as those advanced above having to do with filling the sound-color space, account for most of these choices of sound color, we nevertheless cannot rule out the possibility that the composers had something like a color-inversion operation in mind.

On the other hand, we can hardly expect a composer "unconsciously" to use extensive transposition or inversion of sound colors. These are complex operations that would require, for a sequence of even five sound colors, a purposeful working out. No clear-cut instances of this have been found.

A SPECIAL CASE: STOCKHAUSEN'S *Kontakte*

One of the most analyzed single passages in electronic music is at the precise center of Stockhausen's *Kontakte* (1963).[5] The passage begins (at 17:00.5) with a buzz consisting of a pulse wave having a definite pitch and filtered so as to produce an [aw]-like sound color. Then the pulse frequency is lowered gradually until it falls below the frequency range within which pitch is perceived and the individual pulses of the source are heard. The filter's bandwidth is then narrowed so that it begins to ring at its resonance frequency; what was sound color has now become pitch. The ringing of the filter is then extended longer and longer until it becomes a sustained "pedal" tone upon which a large section of the rest of the composition is superimposed.

Far from contradicting the theory of sound color, this passage is the kind of exception that supports the rule. Pitch and sound color are clearly distinguished at the beginning of the passage. Only when the filter is made to ring is color turned into pitch. At this point, the potential for further sound-color manipulation of the "pitched" filter becomes clear. In fact, later in the work Stockhausen refilters the extension of that sound. Both the dependence of sound color on spectrum envelope and the use of sharply tuned, high-Q filters, such as those employed in most musical instruments to produce pitch, are illustrated in this remarkable passage.

VOCAL MUSIC AND THE SOUND COLOR THEORY

As was indicated at the beginning of this chapter, little independent evidence for the sound-color dimensions and the operations can be

---

[5] A score of this work is available (Stockhausen 1966) and Stockhausen (1962) himself has discussed the passage, concentrating on its temporal aspects.

derived from music for voices. The problem is not a lack of instances of, for example, sound-color changes that suggest the dimensional structure of sound color postulated in Rule 2. Rather, such evidence is weak because the passages can be explained in alternate ways, either as a result of the multidimensional structure of the vowels in a text, or because the transformations of sound color follow other, less abstract regularities as well as those predicted by the theory. Nevertheless, the inverse of this—composing with sound color using the human voice as an instrument—is entirely possible, and a few attempts have been made in that direction. Because the evidence for the theory of sound color in this kind of music is always ambivalent, only a few instances will be cited and no analysis will be attempted.

The phonetic component of Babbitt's *Phonemena* (1977) is based on series of twelve vowels and twenty-four consonants. The succession of phonemes appears to constitute an important aspect of the form of the piece, both as an independent structural element and as a means of articulating the pitch structure. The structure of the piece is clearly quite complex and would require extensive analysis to determine the precise roles of pitch and the phonetic structure. No uses of the sound-color operations specified by Rules 3a and 3b are apparent upon cursory examination of the score[6] or after a few fairly careful hearings of the recordings.

A number of composers, most prominently at the Center for Music Experiment in San Diego, have experimented with extending the uses of the human voice (e.g., Oliveros 1968). Much of their effort has been in the direction of developing new means of articulation that involve unusual source types—growls, clicks, etc. There appears to be little specific concern with developing means of organizing sound color in pieces employing extended vocal techniques.

A remarkable group of female singers calling themselves "Sweet Honey in the Rock" (Barnwell et al. 1982) use sound color in a particularly interesting way. Rooted firmly in the black gospel tradition, these singers exert a control over vocal color that extends that tradition

---

[6] I am grateful to Lynn Webber, the singer who premiered *Phonemena*, for letting me examine her score of the work.

remarkably. The singers vary the "brightness" of their sound, possibly by altering their vocal source characteristics and/or by adjustments that change the effective length of their vocal tracts. Thus, the vowel colors of a song are projected with a "dull" or a "bright" timbre or any of many gradations in between. These kinds of color changes provide examples of "stretching" the sound-color space for which, as in the vowels of children, a normalization process must be brought into play to identify the vowel colors. That such a transformation can be controlled by the singers in this group testifies to the extraordinary flexibility of the vocal apparatus and suggests something of the expressive potential of stretching the sound-color space.

TIMBRE PIECES ORGANIZED IN ALTERNATIVE WAYS

A number of works have been composed in the last two decades that ostensibly do not involve the manipulation of sound color but are strongly concerned with timbre. In these pieces, according to explanations by their composers, regularities other than those postulated in the present study are employed. In some cases, the overtone series is the basis for organization. Pieces so organized will, in general, not control sound color as an independent musical parameter; thus, they appear to be counter-instances to the theory of sound color.

A set of works by Hubert S. Howe, Jr., makes use of the harmonics of a periodic, complex signal as a series that is manipulated in ways analogous to those employed in music based on twelve-tone pitch series. The structures of *Third Study in Timbre* (Howe 1977) and *Improvisation on the Overtone Series* (Howe 1980) are controlled in large measure by the various versions of these series (Howe 1978). The manipulation of the partials in these computer-synthesized works is clear, although the particular sets of partials and their orderings are subtle and to hear them requires repeated listenings. Interestingly, the sound colors in these works are also very clear. At the beginning of *Third Study in Timbre*, for example, each event consists of a changing sound color that, smoothly and audibly, traces the periphery of the sound-color space, [uu oo aw aa ae ee ii]. Sound color is not explicitly controlled, but it nevertheless is strikingly apparent in the sound of the piece. This piece cannot be considered a counterexample to the theory but, rather, an instance in which the perceptual force of sound color is demonstrated in the face of compositional procedures that were concerned with another set of regularities.

Stockhausen's *Stimmung* (1971) is another work based, according to its composer, on the overtone series. The piece is for six singers whose voices are amplified and distributed spatially, by means of a central mixing panel, to speakers placed above and around the audience. The sung pitches of the work are all derived from the harmonics of the B-flat below cello C. The score consists largely of vowels notated in phonetic symbols and includes occasional poetry, "magic words," etc. The sounds of the piece are not intended to be heard as vowels, however. The phonetic symbols are the means of indicating which partial of the sung sound is to be reinforced by adjustment of the singer's vocal tract. The vocal filter in this piece is to be used as a kind of selection mechanism, attenuating all frequencies except that of the selected partial. Nevertheless, vowels are clearly heard throughout the work; indeed, the piece can be heard, "wrongly" perhaps, as a sound-color composition as well as one that manipulates overtone structures.

Boretz's computer piece called *Group Variations* (1974) is primarily about pitch and duration. The timbres in the piece serve to differentiate the contrapuntal lines—the "tunes," as Boretz terms them (slip-cover notes, Boretz 1974). As in most computer music written for the MUSIC programs (Mathews 1969) and modifications of those programs (Howe 1975, 175–248), the waveforms, not the spectrum envelopes, are kept invariant as the pitch changes. This results in the relative-pitch transformation, and sound color varies with pitch. Interestingly, in this piece certain ranges of colors—versions of [ae] or [aa aw] or [oo uu]—are heard as associated with specific contrapuntal lines. The different waveform shapes produce spectra with formant-like peaks, of course, and thus a variety of colors are heard. In this piece, even though the colors vary with pitch, the changes in color *within* single contrapuntal voices are fairly small. On the other hand, the sharp contrasts in sound color *between* the voices suggest that the composer chose waveshapes for each voice that would result in strongly contrasting spectra. The equivalence classes that result are not specific sound colors but, rather, ranges of sound color. Even here, where the basic principles of the computer program steer the composer away from sound-color invariance, choices appear to have been made that recognize that invariance—or, to be exact, very limited sound-color variation—as having an important musical function.

Considerable interest has been aroused by computer techniques for analyzing speech sounds and resynthesizing them with modifications. Dodge's *In Celebration* (1976a), *Speech Songs* (1976b), and *The Story of Our Lives* (1976c) are early examples of the musical application of these techniques.[7] Since speech sounds are essential elements of this music it could be cited, presumably, in support of the theory. Indeed, to the extent that the musical structure in these pieces depends on the vowel color "melodies" of the speech, these pieces are evidence for Rule 1. However, the sound-color structure in most of these cases is given by the original speech that formed the "source" for the composition. The pitches of the original are changed, multiple sources with different pitches are used to excite a single vocal-filter pattern, the temporal course of the speech utterance is modified, the original utterance is mixed with modified versions of itself, etc. The compositional process consists almost entirely of manipulation of the source characteristics.

However, the techniques applied by Dodge and others make possible the manipulation of the F-patterns in analyzed speech. It would be interesting to attempt to perform the sound-color operations of transposition and inversion on preexisting speech sounds, although the calculations would be quite complex. The works by Dodge do not appear to exploit sound color as a musical parameter explicitly, but the techniques he and others have developed suggest a basis for the manipulation of "real" speech and other natural sounds according to the sound-color operations.

---

[7]Lansky's *Six Fantasies on a Poem by Thomas Campion* (1982) also makes elaborate use of computer analysis and resynthesis methods (Lansky and Steiglitz 1981).

SUMMARY AND CONCLUSIONS

In a variety of ways, works of electronic music—especially those that make extensive use of filters—provide considerable evidence that sound color is an independent, musically workable parameter of sound and that it is a function of spectrum envelope. Certain passages in some of these works suggest that their composers hear sound color as if it were organized in a multidimensional perceptual space. Of the dimensions postulated by Rule 2 of the theory, those that seem to be most prominent in these works are ACUTENESS, OPENNESS, and, to a lesser extent, SMALLNESS. The manipulations of sound color in certain passages of electronic music are consistent with the postulated operations of inversion with respect to SMALLNESS and, in one case, of inversion with respect to OPENNESS. Although alternative explanations can be offered for these sound-color transformations, these cases represent some evidence of the use of the operations on sound color in extant works of electronic music.

Experiments with the composition of abstract sound-color "texts" for voice, although providing little independent evidence pertaining to the theory of sound color, do suggest ways that the theory could be applied in vocal composition.

This very selective survey is only the beginning of what could be an extensive study of the use of sound color by composers of electronic music. Clearly, the highly informal and subjective analyses upon which this chapter is based are only preliminary. Analytic methods that preserve a measure of objectivity and can be replicated should be developed for the study of sound color in musical contexts. Perhaps some existing psychoacoustic methods could be adapted for that purpose. Such methods would not only marshal evidence for or against the theory of sound color, they would be likely to result in careful listening to the coloristic aspects of electronic music and they would challenge composers to deal with sound color in interesting ways.

CHAPTER SEVEN

Composition with Sound Color

THE EVIDENCE FROM a variety of scientific fields and from analyses of electronic works by a number of composers strongly supports Rule 1 of the theory of sound color. Sound color can be treated confidently as an independent musical and psychological continuum that is a function of spectrum-envelope maxima caused, typically, by resonances in filters. The evidence also supports (although less strongly) the specification of the dimensions of sound color as stated in Rule 2. For Rules 3a and 3b concerning the operations on sound color, the empirical evidence is weak.

Composers, however, are likely to regard manipulation of sound color by means of operations as the most interesting part of the theory. What composers want from any theory of music are rules by which sound structures may be transformed, and Rules 3a and 3b postulate exactly that. They define sets of transpositions and sets of inversions—options a composer can use to generate different versions of some original set of sound colors. It is not surprising that there has been little direct scientific or musical investigation of these matters: since there has been no theory of sound-color operations, there has been nothing to test scientifically or to apply musically. Recently, however, some compositional work has been done, by the present author and his colleagues and students, that employs the operations on sound color. This final chapter is a description of some of that work.

In this chapter, as in each of the previous ones, a particular method is employed. Taking the theory presented in Chapter Two as given, the implications of that theory for the compositional process are pursued, first generally and then more specifically. Finally, the compositional design of a single extended piece by the author is discussed. This discussion constitutes the final and most substantial evidence for the music usefulness of the sound-color operations.

A single work that applies the theory written by its composer is not, of course, true objective evidence for the theory. However, actual compositions based on the theory show concretely that it has musical relevance. The author's *Colors* is such a composition. In this electronic work, composed and realized in the Computer and Electronic Music Studio at the University of Pittsburgh,[1] sound colors are combined— to paraphrase the remarks of Schoenberg cited at the beginning of the book—not according to the composer's feelings, but according to laws that are comparable in strength and generalizability to the rules for the musical control of pitch.

The compositional, or esthetic, stance of the discussion in this chapter is that of serialism. There is little need to defend this stance; the music that has resulted from serialism, applied strictly or loosely, speaks for itself. In fact, it is likely that far more composers than call themselves serialists are influenced by serial concepts. On the other hand, the approach taken in this chapter need not put off composers who take an "intuitive" approach to composition. Much of what will be dealt with here can be applied directly to music in which structure is determined in an intuitive manner. The serial approach presents, in effect, one extreme of systematic organization that can be sampled to whatever degree a composer wishes.

An important contribution of the theory of atonal music is the development of an unambiguous notation for describing pitch sets, for

---

[1] Preliminary work on *Colors*, supported in part by a grant from the National Endowment for the Arts, was done at the Center for Computer Research in Music and Acoustics of Stanford University in 1979.

designating the operations defined on those sets, and for talking about the consequences of those operations (Forte 1973). In this chapter an attempt is made to develop a similar notation for expressing operations on sound color and to use that notation to describe the effects of the operations on particular collections of sound colors.

CLOSURE, NORMALITY, AND CYCLIC NOTATION

We are concerned in this chapter with certain kinds of sound-color collections.[2] These collections are to sound color what the twelve-tone aggregate is to pitch. The collections are "universes" of sound colors for a particular piece or section of a piece. The difference is that composers use pitch classes that do not belong to the standard twelve pitch-class aggregate (e.g., quarter-tones) only rarely, whereas the selection of a sound-color collection, or "aggregate," is a matter that is open to compositional choice.

Not all collections of sound colors distributed across the sound-color space are amenable to the systematic application of the operations of transposition and inversion. In particular, collections over which the operations are not "closed" are problematic. In non-closed collections an operation will not always transform a color into another color within the collection, but may generate colors that are not in the original collection. The aggregate of twelve equal-tempered pitch classes is closed with respect to the pitch operations of transposition and inversion, but in general, subsets of the aggregate are not: transposition of the trichord (C-C♯-D) up by a semitone drops the C and adds a D♯. Sound-color collections share this lack of closure for sub-

---

[2] *Sound color* is used in this chapter somewhat differently than in the previous chapters. Here what is meant, strictly speaking, is a sound-color class, though not in the sense of *pitch class*, in which octave equivalence is assumed. A sound color in this slightly altered usage is itself a collection of sound colors whose F-patterns cluster around an area on the F_1–F_2 plane. Following the practice of phoneticians, who speak of vowels rather than vowel classes, we shall speak simply of sound colors, with the implicit assumption that they represent categories of different filter configurations.

sets, but a more serious kind of non-closure is brought about if the complete collection, unlike the twelve-tone pitch set, is not closed with respect to one or more of the operations.

Certain distributions of sound colors in the sound-color space are closed with respect to most of the operations. On the other hand, there are very few sets in which *all* transpositions and inversions are closed. Transposition with respect to LAXNESS is the most problematic because, as will be shown below, sets in which that operation is closed are awkward in other ways. For this reason, it is useful to require somewhat less than complete closure for all the operations. Let us define a *normal* sound-color collection as one that is closed for transposition and inversion with respect to ACUTENESS, OPENNESS, and SMALLNESS. In such collections, LAXNESS transposition may not be closed, but the regular behaviors of the other operations provide a considerable range of transformational possibilities.

Let us consider as a first example of a normal collection the particularly interesting set of nine sound colors distributed in three rows of three colors each across the sound-color space. That space, it will be recalled, is a perceptual space that can be represented as a plane whose vertical axis is ACUTENESS and whose horizontal axis is OPENNESS (see Figure 37). Included at the four corners of the space are the upper and lower extreme values of each of those two dimensions ([ii], [uu], [ae], and [aw]), medium or neutral colors on each of the axes at the upper and lower extremes of the other axis ([ee], [oo], [oe], and [aa]), and the neutral color itself. The SMALLNESS dimension is also represented by extreme upper and lower values ([ae] and [uu], respectively), and colors that are neutral with respect to it ([ii], [ne], and [aw]). There are also colors with medium-high and medium-low SMALLNESS ([ee aa] and [oe oo], respectively). In this set there is only a single LAX sound color—the neutral color [ne]; all the others are equally non-LAX. The only LAXNESS transposition that is closed transforms all those non-LAX colors into the single neutral color. A LAXNESS transposition any smaller in degree generates colors corresponding to the short vowels that are not within the collection. This collection illustrates well the common case in which LAXNESS transposition presents a certain anomaly but the other operations function quite regularly.

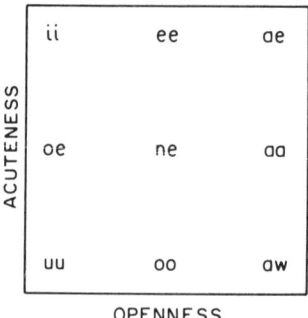

FIGURE 37: *Nine-element sound-color collection in the sound-color space. OPENNESS-horizontal axis; ACUTENESS-vertical axis.*

As a consequence of the property of normality, the operations of transposition and inversion with respect to ACUTENESS, OPENNESS, and SMALLNESS can be specified simply and precisely with a notation borrowed from group theory, called *cyclic notation*. In this notation, the elements of a collection are organized into subsets in a certain way for each operation. In the nine-color collection of Figure 37, for example, the operation of transposition with respect to ACUTENESS can be specified as (uu oe ii)(oo ne ee)(aw aa ae). Each subset is construed as a cycle, with the number of moves through the cycle corresponding to the degree or "size" of the transposition. Suppose we wished to transpose some colors from the collection one step in the direction of increasing ACUTENESS. We would find each of the original colors in the cyclic representation of the collection and transform it into the color one step to its right—e.g., [aa] becomes [ae]. If in the process we reach a right parenthesis, we jump to the left parenthesis and take the leftmost color of the cycle—e.g., [ee] becomes [oo]. A two-step transposition in the same direction transforms a color into the color two steps to its right (with the same provision when reaching the right parenthesis)—e.g., [oe] becomes [uu]. Transposition in the direction of decreasing ACUTENESS calls for moving to the left within each cycle. Each of the operations is defined with this notation in Table 1.

One advantage of specifying operations in this way is the clarity with which the number and type of unique operations are displayed.

TABLE 1: *The operations defined in cyclic notation on the normal nine-color collection*

transposition	
ACUTENESS	(uu oe ii)(oo ne ee)(aw aa ae)
OPENNESS	(uu oo aw)(oe ne aa)(ii ee ae)
SMALLNESS	(uu ne ae)(oo aa)(oe ee)(aw)(ii)
inversion	
ACUTENESS	(uu ii)(oo ee)(aw ae)(oe)(aa)(ne)
OPENNESS	(uu aw)(oe aa)(ii ae)(oo)(ee)(ne)
SMALLNESS	(uu ae)(oo aa)(oe ee)(aw)(ii)(ne)

The notation makes it obvious, for example, that transposition with respect to ACUTENESS of three steps in either direction is the same as no operation at all, and that transposition of two steps in one direction is equivalent to transposition of one step in the opposite direction. Similarly, the notation shows that OPENNESS and ACUTENESS transposition have only three unique forms. Inversions of inversions in the same dimension produce the original set.

Transposition in SMALLNESS is a special case. We can preserve closure somewhat artificially by defining cycles of different lengths—the (uu ne ae) cycle has three elements whereas the (oo aa) and (oe ee) cycles have only two. A total of six forms, the prime and five transpositions, are produced by successive applications of that operation. Since the point at which wrap-around occurs depends on the length of the cycle, SMALLNESS transposition can result in drastic alterations of a sound-color set, which threatens the perceptual invariance claimed for that operation.

The geometric equivalent to normalcy in a sound-color collection is symmetry about the neutral equal-value contours of each dimension. Considered in this way, it becomes obvious why LAXNESS is not involved in the question of normalcy; there is no neutral contour for that dimension. Without symmetry, inversion clearly will produce colors outside the collection and transposition can only be defined with closure in a geometrically and psychoacoustically unrealistic way.

PROPERTIES OF SOME NORMAL SOUND-COLOR COLLECTIONS

What other collections of sound colors are normal, and what are their properties? It can be verified that collections with four, nine, sixteen, and twenty-five sound colors arranged in squares in the sound-color space are all normal. Moreover, eight- and twenty-four-element collections—derived from squares of nine and twenty-five elements, respectively, with the neutral middle element removed—are also normal. In fact, all square collections of colors are normal, but collections larger than twenty-five would be too large for listeners to be able to identify the individual colors in even the most ideal musical context. Indeed, the twenty-four- and twenty-five-color collections may be problematical.[3] The square collection of four colors, on the other hand, is rather impoverished and is missing the neutral color. The sixteen-color collection, richer than the nine-color collection, is probably an appropriate basis for composition. However, even sixteen sound colors may present the listener with problems of identification. The situation can be compared to that of the vowel systems of languages; only a few—Swedish, for example—have as many as sixteen vowels distinguishable on the basis of resonance frequencies alone.[4]

Twelve-Color Collections

Collections of twelve colors present the possibility of establishing direct correspondences between sound colors and pitch classes. A nonsymmetrical arrangement of twelve sound colors would leave opera-

---

[3] As was indicated in Chapter Four, the number of colors that can be distinguished from one another may be considerably larger than twenty-four or twenty-five. We are concerned here, however, with the number of sound colors in a *collection* that can be recognized or identified—a more demanding task than simple discrimination, but one that ordinarily would be required for the perception of a musical structure based on sound color.

[4] Certain languages, Thai for example, have very large numbers of vowels, but in those cases the vowel features include linguistic tones, generated by the laryngeal source, not the filter. Swedish is usually considered to have twenty vowels (Fant 1959), and tone is not among its vowel features.

tions with respect to SMALLNESS ill defined, but there is at least one promising symmetrical twelve-color collection (displayed in Figure 38). It is normal; the operations of transposition and inversion with respect to ACUTENESS, OPENNESS, and SMALLNESS—properly interpreted—are closed. (The operations for this normal twelve-color collection are defined in Table 2.) The inner four colors, when operating with respect to ACUTENESS and OPENNESS and when inverting with respect to SMALLNESS, are transformed only into each other. Only transposition with respect to SMALLNESS maps the inner four colors into the less LAX colors in the periphery of the space. This limitation of "mixing" among the inner four colors and the outer eight could be of considerable importance in large-scale organizations based on this twelve-color collection.

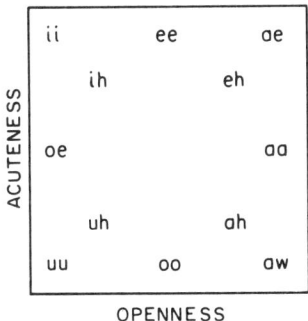

FIGURE 38: *The normal twelve-color collection.*

Collections with LAXNESS Closure

Another class of symmetrical sound-color collections permits transposition with respect to LAXNESS. One of these color collections is illustrated in Figure 39. Like the twelve-color collection, this eight-color collection, a square within a square, tends to isolate the inner from the outer set of colors except for transposition with respect to SMALLNESS. In this collection, however, LAXNESS transposition is well defined and also provides a mapping between the inner and outer groups of four colors. If the outer colors are [ii ae aw uu] and the analogous inner colors are [ih eh ah uh], then transposition in LAXNESS can be defined

TABLE 2: *The operations defined
on the normal twelve-color collection*

transposition	
ACUTENESS	(uu oe ii)(uh ih)(oo ee)(ah eh)(aw aa ae)
OPENNESS	(uu oo aw)(uh ah)(oe aa)(ih eh)(ii ee ae)
SMALLNESS	(uu uh eh ae)(oo ah aa)(oe ih ee)(aw)(ii)
inversion	
ACUTENESS	(uu ii)(uh ih)(oo ee)(ah eh)(aw ae)(oe)(aa)
OPENNESS	(uu aw)(uh ah)(oe aa)(ih eh)(ii ae)(oo)(ee)
SMALLNESS	(uu ae)(uh eh)(oo aa)(oe ee)(aw)(ah)(ih)(ii)

as (ii ih)(ae eh)(aw ah)(uu uh). A sixteen-color collection with eight colors in both inner and outer "shells" is another example of this class of color sets.[5]

Compositional Implications of Closure and Non-Closure

Clearly, the sound-color space can be organized in a variety of ways, each with its strengths and weaknesses. The sound-color collections that are normal appear to be the most useful for composers who value transformations that form mathematical groups. It is perhaps unnecessary to add that the theory, in the form presented in Chapter Two, in no way depends on such considerations. Composers who prefer more intuitive compositional methods can apply the operations of inversion and transposition to sound-color arrays built from a structure using quite different premises—"gesture," preexisting forms, or text setting, for example.

Both transposition in LAXNESS and "stretching" of the sound-color space—two typically non-closed operations—may be regarded as changes in the sound-color space itself. Such changes cannot func-

---

[5] If we assume, as a first approximation, that our sensitivity to changes in sound color is approximately evenly distributed across the color space, then the sound-color collections that are closed with respect to LAXNESS can be said to violate psychoacoustic regularity; LAXNESS-closed collections have more colors concentrated in the center of the space than at the periphery.

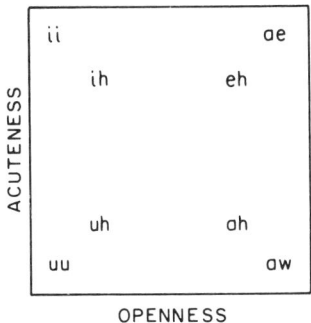

FIGURE 39: *An eight-color collection that permits LAXNESS transposition.*

tion as part of regular, or serial, sound-color structures. They may be considered analogous to operations that are entirely extraneous to sound color—e.g., changes in pitch or dynamics. It follows that a sound-color collection can be stretched or collapsed in on itself without affecting the other properties of the collection—for example, its normalcy—at all. Just as pitch, loudness, or changes in other source characteristics can be used to point up sound-color structures, the inherently non-closed sound-color transformations of stretching and, in most collections, LAXNESS transposition may be applied quite independently of other sound-color operations.

MULTIPLE OPERATIONS ON
THE NORMAL, NINE-ELEMENT COLLECTION

The nine-element collection shown in Figure 37 has been investigated more than those of any other size or distribution in the sound-color space. The investigation has focused on that collection taken as an ordered series of sound colors, although certain of the results apply equally well to the unordered collection.[6] It is often easier to follow the results of operations in ordered color sets, however, than in unordered sets. Let us consider, therefore, the consequences of the operations,

---

[6] If, for example, a normal collection is treated as an unordered aggregate, then the effects of the operations on unordered subsets can be investigated. Lovendusky and Slawson have embarked on such a study of the nine-element, normal collection, both theoretically (in preparation) and as applied to composition.

applied both singly and in combination, to the nine-element normal collection treated as an ordered row of sound colors.

A Convention for Naming the Operations

As a convenience in discussing multiple operations we shall establish a convention and shorthand notation. The original or prime form of any ordering of the total collection will be designated P. Transpositions will be identified by the letter T followed by an abbreviation of the dimension—O for OPENNESS, A for ACUTENESS, S for SMALLNESS—and finally by a number indicating the degree of the transposition. Thus, TA_2 is to be interpreted as a two-step transposition in the direction of increasing ACUTENESS. It follows that TO_0, TA_0, and TS_0 are all equal to each other and to P.

Since inversion is a binary operation, we need only specify the dimension with respect to which the inversion is to be carried out. Adopting the same abbreviations for the dimension names, we can designate the three varieties of inversion as IO, IA, and IS.[7]

The order of application of combinations of operations will be designated by parentheses, with the inner operation construed as having been carried out first and the outer operation last. Thus $TS_3(IO(IA))$ would designate the inversion of a series with respect to ACUTENESS, inversion of the resulting series with respect to OPENNESS, and finally three degreees of SMALLNESS transposition of the result of the successive inversions.[8]

---

[7]Since the nine-element set is not closed with respect to LAXNESS, that dimension is ignored in the following discussion.

[8]The order of the six SMALLNESS transpositions is derived by stepping through the groups defined in Table 1 from left to right. Let us choose an ordering of the complete nine-element collection as an example. The SMALLNESS-transpositions would be as follows:

$P =$	TS_0:	oo	ee	uu	aw	ii	ae	oe	aa	ne
	TS_1:	aa	oe	ne	aw	ii	uu	ee	oo	ae
	TS_2:	oo	ee	ae	aw	ii	ne	oe	aa	uu
	TS_3:	aa	oe	uu	aw	ii	ae	ee	oo	ne
	TS_4:	oo	ee	ne	aw	ii	uu	oe	aa	ae
	TS_5:	aa	oe	ae	aw	ii	ne	ee	oo	uu.

Rules for Combining Operations

Operations on series of sound colors, in general, permute the order of the colors in the series. It follows that two or more successive operations will produce additional reorderings. Not all the combinations of operations result in unique series, however. Certain operations, when combined, produce the same orderings as other operations or combinations of operations. Either by reasoning from the definitions of the operations or by simply performing the specified operations, it can be shown that the following propositions hold for the normal, nine-color series.

Proposition 1: For all successions of transpositions with respect to the same dimension, X,

$$TXm(TXn) = TXr, \text{ where } r = m + n \text{ (mod 3 for OPENNESS and ACUTENESS; mod 6 for SMALLNESS)}.$$

Example: TA2 of TA1 is the same as TA0 (or P).

Proposition 2: For transpositions of levels m and n equal to 0, 1, or 2,

$$TAm(TOn) = TOn(TAm).$$

Example: TA2 of TO1 is the same as TO1 of TA2.

Proposition 3: For OPENNESS and ACUTENESS inversions, IY, and for all OPENNESS and ACUTENESS transpositions, TXm ($m = 0, 1, $ or 2),

$$TXm(IY) = IY(TXm)$$

Example: TA1 of IO is the same as IO of TA1.

Proposition 4: For all pairs of inversions with respect to the same dimension, X = O, A, or S,

$$IX(IX) = 0 \text{ (the null operation)}.$$

Proposition 5: For inversions with respect to ACUTENESS and OPENNESS,

$$IA(IO) = IO(IA).$$

Proposition 6: For inversions IS, IO, and IA,

$$IS(IO) = IA(IS),$$

and

$$IS(IA) = IO(IS).$$

Propositions 2, 3, and 5 imply that with respect to OPENNESS and ACUTENESS, inversions and transpositions may be performed in any order without changing the result. This is not the case for SMALLNESS, where the order of operations is critical. For example, $IS(TAm)$ is not, in general, the same as $TAm(IS)$.

Proposition 1 implies that any succession of transpositions in one dimension can be reduced to a single transposition in that dimension. We can deduce from Propositions 1 and 2 that any mixed succession of transpositions in OPENNESS and ACUTENESS can be reduced to a pair of such transpositions. For example, $TA2(TO1(TA2(TO2)))$ can be reduced to $TA2(TA2(TO1(TO2)))$, to $TA1(TO0)$, and finally simply to $TA1$.

Successions involving SMALLNESS transpositions are less easily reduced, since there are no simple relationships between the various SMALLNESS transpositions and operations with respect to the other dimensions. However, three of the SMALLNESS transpositions—those of odd degree: $TS1$, $TS3$, and $TS5$—may be construed to be maximally similar to SMALLNESS inversion.[9] That is to say, the SMALLNESS inversion of a given ordering of the nine-color series differs from the odd SMALLNESS transpositions of that series in only two order posi-

---

[9] *Maximal similarity* in the present context is to be distinguished from the term as applied to pitch-class sets. The similarity of ordered sound-color series is calculated here with reference to the number of exchanges among elements of the series, not to the interval content of an unordered set, as in Forte's theory of set complexes (1973) (cf. Morris 1979–1980).

tions—the minimum by which two distinct series can differ. Clearly, similarity relations among operations on sound-color series in general and among particular sound-color series are a rich area for investigation, but little formal consideration has been given to such relations and the topic will not be developed further here.

Although these propositions are derived explicitly for the normal, nine-color collection taken as an ordered series, similar sets of propositions can be shown to hold for the normal collections of other cardinalities that are perfect squares and, it follows, for operations on ordered or unordered subsets of those collections. For example, propositions 1, 2, and 3 would require only an increase in the number of possible degrees of transposition to four in order to be applicable to the normal sixteen-color collection. Most normal collections of cardinalities that are not perfect squares do not have analogies to the propositions dealing with transposition, because transposition in those collections is defined by cycles of differing length. The eight-color collection that permits LAXNESS transposition is an exception. Propositions 1, 2, and 3 need only be reduced to degree two to hold—although rather trivially—for that collection. In all normal collections the propositions dealing with inversion hold without alteration.

The Unique Inversions on the Normal Nine-Color Collection

The six propositions about combinations of operations can be used to show that only eight unique series, including the prime, can be produced from combinations of inversions alone. Table 3 lists the eight combined operations and the series that result when they are applied to a particular ordering of the nine colors.

The proof of this assertion is not difficult. We can argue first of all that any combination of four or more inversions can be reduced to three or fewer by interchanging the order of the inversions (Proposition 5) until pairs of repeated inversions with respect to the same dimension are consecutive, and then eliminating the pairs (Proposition 4). By similar reasoning, it can be shown that IO(IS(IA)) is the same as IS and that IS(IA(IO)), IO(IA(IS)), and IA(IO(IS)) are equal to each other and to IS(IO(IA)). Other identities that follow directly

TABLE 3: *The eight unique combinations of inversion operations on the nine-color collection, and the orderings that result from them*

OPERATIONS	SERIES
P	oo ee uu aw ii ae oe aa ne
IO	oo ee aw uu ae ii aa oe ne
IA	ee oo ii ae uu aw oe aa ne
IS	aa oe ae aw ii uu ee oo ne
IO(IA)	ee oo ae ii aw uu aa oe ne
IS(IO)	aa oe aw ae uu ii oo ee ne
IS(IA)	oe aa ii uu ae aw ee oo ne
IS(IO(IA))	oe aa uu ii aw ae oo ee ne

from Propositions 4, 5, and 6 establish that IA(IO)), IA(IS), and IO(IS) are each equal to one of the seven primary inversion combinations. The other possible inversion combinations similarly can be reduced to one of the seven. Although this reduction of the number of inversion combinations to eight (including the prime) has been worked out explicitly only for the normal nine-color collection, it applies as well to all normal collections. This follows from the fact that the propositions dealing with inversion, upon which the reduction is based, apply to all normal collections.

The Unique Transpositions on the Normal Nine-Color Collection

Propositions 1, 2, and 3 depend on the number of elements in the collections to which they are to apply; it follows that the number of unique transpositions in any normal collection is a function of its cardinality. However, unless reduced by identities derived from those propositions, as in the case of the inversions, the number of unique orderings generated by combinations of transpositions is bounded only by the set of all possible orderings—in the case of the nine-element array, 9! or 362,880. In non-square collections—for which, as we have seen, there are no simple versions of the relevant propositions—we can guess that the operations may indeed generate the total number of unique orderings. If they do, then such collections are of little interest because any given ordering is as derivable as any other. In square sets, on the other hand, to which the propositions apply, the operations generate a subset—possibly in some cases a small, and therefore musically interesting, subset—of the possible orderings.

For concreteness, let us apply these observations to the normal nine-color set. Since there are three degrees of transposition in ACUTENESS and OPENNESS, the total number of transpositions involving only those dimensions is nine, including the prime form. We can draw this conclusion because all combinations of three or more such transpositions can be reduced to two or fewer by appropriate application of Propositions 1 and 2. For example, $TA_1(TO_1(TA_2))$ can be transformed into $TA_1(TA_2(TO_1))$ (Proposition 2), then into $TA_0(TO_1))$ (Proposition 1), and finally into TO_1. The nine unique combinations of ACUTENESS and OPENNESS transpositions in the square nine-color set are P, TA_1, TA_2, TO_1, TO_2, $TA_1(TO_1)$, $TA_1(TO_2)$, $TA_2(TO_1)$, and $TA_2(TO_2)$. By similar reasoning it can be verified that *in all square collections* the number of unique transpositions involving only ACUTENESS and OPENNESS is equal to the cardinality of those sets.[10]

Combinations of Transpositions and Inversions

The only general rule for the combination of the operations of transposition and inversion is Proposition 3, which asserts the commutativity of those two operations with respect to the dimensions of ACUTENESS and OPENNESS. It follows from that proposition and the totals of

---

[10] I have purposely avoided encumbering the present discussion with rigorous use of concepts from group theory. Not much of substance, at this point, would be gained by such rigor. Nevertheless, it is clear that the transposition and inversion operations generate a subgroup of the nine-element normal array. The cardinalities of this subgroup and those of other collections are a critical issue that requires further research, where no doubt a more rigorous approach will be required.

four unique inversions and nine unique transpositions involving only ACUTENESS and OPENNESS that there are 36 unique combinations of inversions and transpositions with respect to the dimensions of ACUTENESS and OPENNESS in the normal nine-color set. Because of the irregular nature of the operations involving SMALLNESS, they do not commute with the ACUTENESS and OPENNESS operations.

Although SMALLNESS inversion does not commute with transposition with respect to ACUTENESS and OPENNESS, it appears that all eight inversions, including those involving SMALLNESS, can be combined freely with ACUTENESS and OPENNESS transpositions without enlarging the number of unique orderings resulting from application of the eight times nine operations. That is to say, the SMALLNESS inversion of one of the nine transpositions is not, in general, the same as that transposition of the SMALLNESS inversion, but both orderings will be among the 72 generated by the combined operations. Like the 48 variants of a twelve-pitch-class set, these 144 versions (including the retrogrades) of the nine-element color series form a closed system in spite of the non-commutativity of certain of the operations.

Transposition with respect to SMALLNESS, however, generates a large number of orderings not among the 144. At present, the number of unique combinations of operations that involve SMALLNESS is not known.

Subsets of the Normal Eight- and Nine-Color Collections

The properties of the unordered four-color subsets of the square, eight-color collection have been investigated (Lovendusky and Slawson, in preparation). Such sets are analogous to hexachords in twelve-tone theory; a "tetra-color" and its complement in the eight-color collection share many characteristics; for example, they have the same distribution of invariances under the eight inversion operations. In general, the inversion of a set may be equivalent to the original set or it may include a color or colors not in the original set. For each of the eight inversions, the number of colors shared by the original set and its inversion were calculated. The resulting eight numbers make up an "inversional invariance vector (IIV)."[11] All tetra-colors have the same IIV as their complements. For example, the eight inversions of the set [uu oo aw ae] share with the original set the following numbers of colors:

IO	IA	IS	IA(IO)	IS(IO)	IS(IA)	IS(IA(IO))
3	2	3	2	2	2	2

The complement of that set, [oe aa ii ee], has exactly the same pattern of invariances under each of the inversion operations. In this pair of sets none of the operations transforms a set onto itself—there are no entries of 4 in the IIV. On the other hand, the set [oo oe ee aa] and its complements, [uu aw ii ae], are maximally invariant: all the inversions transform each of these sets into itself. These observations apply equally well to the nine-color normal collection if complementation is defined as excluding the neutral sound color—a justifiable redefinition, because the neutral color always inverts into itself.

THE COMPOSITION *COLORS*

The foregoing brief, informal study of sound-color collections and sound-color operations could be expanded in a number of directions. Most, if not all, of the questions that composer-theorists have asked about pitch-class sets and the well-known pitch operations could be rephrased into questions about sound-color sets and the color operations. However, further theoretical developments probably should await compositional experience with pieces in which sound color is organized according to the theory as it stands at present. A start on such experience is provided by the author's *Colors*. Let us turn to a discussion of that piece and the planning that preceded its composition.[12]

---

[11] The IIV should not be confused with the interval vector calculated for pitch-class sets. The IIV in no way measures the intervals between sound colors, but only invariances under the various inversions.

[12] An earlier version of portions of this discussion appears in Slawson (1982).

Colors is a quadraphonic work for tape alone in the form of eleven variations. It was realized with the Arp 2500 and Buchla 100 synthesizers under computer control in the Computer and Electronic Music Studio of the University of Pittsburgh during the autumn of 1980. It was completed and presented for the first time at the University of Pittsburgh in January 1981.

The Overall Form of *Colors*

The main unifying feature of the variations that make up *Colors* is a structure of sound color synthesized by means of a pattern of changing frequencies of two-formant filters.[13] In each variation, the sound-color structure is more or less the same, whereas the source changes in character from variation to variation. Table 4 summarizes certain general features of the variations.

In the first three variations, a section of the work entitled "The Landscape," the color structure is presented in a straightforward manner as non-varying colors. The sound sources in those variations are a systematic succession and combination of pitched sounds, noise, and frequency-modulated sounds that serve to differentiate and emphasize aspects of the sound-color structure.

In Variations 4 through 7, subtitled "Motions," the same color structure is expressed as continuously changing, diphthong-like sound colors in which the durations of individual colors are less differentiated than they are in the first three variations. In this second group of variations, like the first, the "variation" is largely in the character of the sources. In general, the sources become harsher and louder from Variations 4 through 6. Variation 4 is very slow; Variation 5 is significantly faster, and Variation 6 is faster yet and highly compressed. The seventh variation has the same compressed character as Variation 6, but the pitches are distinct and they refer back to those of the first three variations.

---

[13] The colors were synthesized by computer control of two Arp resonance filters connected in series. The sources that excited the filter network were generated electronically, in part with the Buchla and in part with the Arp. Additional details of the actual synthesis methods are given in the Appendix.

TABLE 4: *General Characteristics of* Colors

VARIATION NUMBER	DURATION (secs.)	SOURCE CHARACTER	COLOR CHARACTER
1	111	pitch/noise[a]	fixed
2	111	AM[b]/noise	fixed
3	111	noise/pitch	fixed
4	170	noise/pitch	varying
5	68	FM[c]/noise	varying
6	17	FM	varying
7	17	pitch	varying
8	123	pulsed/noise	mixed[d]
9	123	pulsed/FM	mixed
10	123	pulsed/noise	mixed
11	116	pitch/noise	fixed

[a]When two source characters are cited, the order in the table indicates the prominence of each or their order of occurrence.
[b]Amplitude modulated.
[c]Frequency modulated.
[d]Varying or ramp color changes in one or more musical strands; fixed or step color successions in the remaining strands.

The next three variations, "Events and Continuities," feature a percussive, pulsating source and a mixture of patterns of fixed and changing colors. The pitch structure that emerges over this group is a transformation of that of the first set of three variations, emphasizing different features of the color structure.

The final variation, "A Return," presents the steady-state color structures of the first group of variations, now "set" with the pitches that have been anticipated only in part in Variations 8 through 10.

The Basic Sound Color Materials

The sound-color set of *Colors* is the square nine-color set that has been discussed in some detail above. No "stretching" of the space toward colors corresponding to the vowels of women and children are included in *Colors*, nor are LAXNESS transpositions. The sound-color structure is derived by transformations on a single ordering of the nine-color set.

Properties of the Ordered Series of *Colors*

The prime order of the color series in *Colors* is [oo (as in "mote"), ee ("hate"), uu ("boot"), aw ("ought"), ii ("keep"), ae ("bad"), oe (German "böse"), aa ("lot"), ne (the "e" in "rated")]. This order of the nine colors is presented graphically in Figure 40. The basic series can be said to emphasize opposites in the succession of colors. The first and second colors, [oo ee], are neutral in OPENNESS, but they are low and high, respectively, in ACUTENESS. The next disjunct pair of colors, [uu aw], are both low in ACUTENESS, providing a contrast in that dimension with the second color, but they are opposite in OPENNESS. In contrast to the second pair, the next disjunct pair, [ii ae], are both high in ACUTENESS and opposite in OPENNESS. The seventh and eighth colors, [oe aa], both have neutral ACUTENESS and, like the second and third pairs, they are opposite in OPENNESS. The final color is the neutral color, [ne], in the center of the color space.

The last three colors, taken as a group, represent the only instance in the series of a sequence of more than two consecutive colors that have the same values on any single dimension. This relationship among the last three colors provides an exception to the strong contrasts between successive disjunct pairs over the first six colors, and as such it is a perceptually distinctive feature of the series. Since the three also are neutral in ACUTENESS, they complete a general tendency for the series to move from strong to less strong contrast from its beginning to its end.

Operations on the Series

As an aid to compositional planning for *Colors*, matrices were assembled that displayed the 144 unique orders of the series resulting from all eight possible inversions and from the transpositions with respect

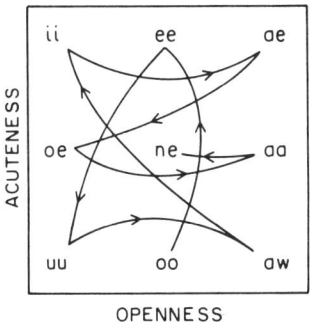

FIGURE 40: *Prime order of the sound-color series in* Colors.

to ACUTENESS and OPENNESS and their retrogrades. These matrices—obviously analogous to the pitch-class tables commonly used in the composition of twelve-tone music—are reproduced in Table 5. Direct and inverted forms are read from left to right and from top to bottom; retrograde and retrograde inverted forms, from right to left and bottom to top. The order of inversions is from right to left and bottom to top; e.g., IS∗IA is the SMALLNESS inversion of the ACUTENESS inversion. The numbers, modulo 3, indicate the degree of transposition of the first element in each series that will produce the succeeding elements. The "threes" position—00, 10, 20—represents transpositions in ACUTENESS; the "ones" position—01, 02, 03, transpositions in OPENNESS.

The four matrices represent only a portion of the total number of series that are generated by the operations in all their combinations. The total of eight inversions of the prime form—the top rows and the left-most columns of each matrix—include all the unique series generated by combinations of inversions alone. The rest of the matrices list the possible transpositions of the eight inversions with respect to OPENNESS and ACUTENESS. SMALLNESS transposition, due to its relatively complex interaction with the other operations, was not used in *Colors* and is not included in the matrices. By convention, all inversions are carried out first, followed by transposition. However, reversing the order of those operations generates only orderings already present in the matrices.

TABLE 5: *The color series of Slawson's* Colors *transformed by the eight inversions followed by the nine transpositions with respect to* ACUTENESS *and* OPENNESS

DIRECT FORMS

		00	20	02	01	22	21	12	11	10
I	00	oo	ee	uu	aw	ii	ae	oe	aa	ne
O	20	ee	ne	ii	ae	oe	aa	uu	aw	oo
	01	aw	ae	oo	uu	ee	ii	ne	oe	aa
	02	uu	ii	aw	oo	ae	ee	aa	ne	oe
	21	ae	aa	ee	ii	ne	oe	oo	uu	aw
	22	ii	oe	ae	ee	aa	ne	aw	oo	uu
	11	aa	aw	ne	oe	oo	uu	ee	ii	ae
	12	oe	uu	aa	ne	aw	oo	ae	ee	ii
	10	ne	oo	oe	aa	uu	aw	ii	ae	ee

IS

		00	01	10	20	11	21	12	22	02
I	00	aa	oe	ae	aw	ii	uu	ee	oo	ne
S	01	oe	ne	ii	uu	ee	oo	ae	aw	aa
*	20	aw	uu	aa	ae	oe	ii	ne	ee	oo
I	10	ae	ii	aw	aa	uu	oe	oh	ne	ee
O	21	uu	oo	oe	ii	ne	ee	aa	ae	aw
	11	ii	ee	uu	oe	oo	ne	aw	aa	ae
	22	oo	aw	ne	ee	aa	ae	oe	ii	uu
	12	ee	ae	oo	ne	aw	aa	uu	oe	ii
	02	ne	aa	ee	oo	ae	aw	ii	uu	oe

IA

		00	10	02	01	12	11	22	21	20
I	00	ee	oo	ii	ae	uu	aw	oe	aa	ne
A	10	oo	ne	uu	aw	oe	aa	ii	ae	ee
*	01	ae	aw	ee	ii	oo	uu	ne	oe	aa
I	02	ii	uu	ae	ee	aw	oo	aa	ne	oe
O	11	aw	aa	oo	uu	ne	oe	ee	ii	ae
	12	uu	oe	aw	oo	aa	ne	ae	ee	ii
	21	aa	ae	ne	oe	ee	ii	oo	uu	aw
	22	oe	ii	aa	ne	ae	ee	aw	oo	uu
	20	ne	ee	oe	aa	ii	ae	uu	aw	oo

IS*IA

		00	02	10	20	12	22	11	21	01
I	00	oe	aa	ii	uu	ae	aw	ee	oo	ne
S	02	aa	ne	ae	aw	ee	oo	ii	uu	oe
*	20	uu	aw	oe	ii	aa	ae	ne	ee	oo
I	10	ii	ae	uu	oe	aw	aa	oo	ne	ee
A	22	aw	oo	aa	ae	ne	ee	oe	ii	uu
*	12	ae	ee	aw	aa	oo	ne	uu	oe	ii
I	21	oo	uu	ne	ee	oe	ii	aa	ae	aw
O	11	ee	ii	oo	ne	uu	oe	aw	aa	ae
	01	ne	oe	ee	oo	ii	uu	ae	aw	aa

Color Combinatoriality

The sound-color "theme" of the eleven variations was constructed by an adaptation to sound color of *combinatoriality*, the technique often used to structure pitch in serial music. This technique was invented by Schoenberg and developed further by Babbitt (1955) and Starr and Morris (1977–1978).

Combinatoriality, as applied to pitch organization, involves the presentation of two or more transformations of a twelve-tone row linearly in separate contrapuntal voices, subject to the condition that all twelve tones must be present in the vertical structure before any tone is repeated. In the extension of the technique developed and formalized by Starr and Morris, the vertical "aggregates" of the twelve pitch classes can be formed quite freely from any of the contemporaneous rows as long as order within the linear rows is preserved. For example, in a five-row structure the first aggregate might be made up of five pitch classes from the first row, two from the second, none at all from the third and fourth, and four from the fifth row. Such an arrangement provides a great deal of contrapuntal flexibility, while preserving the kind of integration of pitch materials that makes the twelve-tone method attractive in the first place.

The basic idea of combinatoriality can be translated quite directly into the structuring of sound color. The prime set as displayed in Figure 40 cannot be combined simply with its OPENNESS inversion, because that operation in this set results in a permutation of consecutive pairs, leaving colors in nearly the same positions in both series. But that same property of the set ensures that the prime form can be combined easily with the retrograde of the OPENNESS inversion. The first four colors of the original series and the first five colors of the retrograde of the OPENNESS inversion can be taken as the first of two aggregates of nine colors, with the remaining colors from each of the series forming the second. This combinatorial matrix is illustrated in Table 6a. The swapping operation described by Starr and Morris can be applied to this matrix to produce another, shown in Table 6b, with quite different contrapuntal possibilities.

RECORDED EXAMPLES 8A, 8B

TABLE 6: *Two-series, two-aggregate combinatorial sound-color matrices. The first matrix (a) can be used to derive the second (b), by means of "swapping."*

		AGGREGATE 1	AGGREGATE 2
(a)	Prime	[oo ee uu aw	ii ae oe aa ne]
	Retro. (IO)	[ne oe aa ii ae	uu aw ee oo]
(b)	Prime	[oo ee	uu aw ii ae oe aa ne]
	Retro. (IO)	[ne oe aa ii ae uu aw	ee oo]

These combinations of the two versions of the basic series are interesting because of the possibilities they present for relating pairs across aggregate boundaries. For example, in the first matrix, the [ii ae] of the first aggregate can be associated—by means of, say, similar pitch registers or rhythmic patterns—with the same pair in the second aggregate. Similarly, the temporal juxtaposition of the pairs [uu aw] in both aggregates can be emphasized by manipulation of the non-coloristic (source) parameters of the sounds.

Combinatorial Sound-Color Structure in Colors

The sound-color structure on which *Colors* is based is a considerable extension of the basic idea of color combinatoriality as presented above. Each half of the bipartite basic structure (the "theme") is a single combinatorial matrix consisting of seven contrapuntal voices with one or two linear versions of the series in each voice. Table 7 shows that combinatorial matrix. Brackets enclose linear forms of the series, omitting the indicators of inversion and transposition and representing the levels of transposition as in the matrices in Table 5, e.g., RAO01 means Retro.(IA(IO(TO1))). Relative durations of each aggregate are given, the first number applying to the first half of the basic structure, the second to the second half (when the matrix is presented

TABLE 7: *Color combinatorial matrix of* Colors

	A1, DUR. 4/4.5	A2, DUR. 5/2.5	A3, DUR. 3/7	A4, DUR 6/2
S1	[P:oo ee uu aw ii ae oe	aa ne]		[P:oo ee uu
S2		[AOo2:ii uu ee ae	oo aw	ne aa
S3			[RAO10:ee ii ae oe	
S4	[10:ne	oo oe	aa uu	aw ii
S5				
S6				[RA11:ae
S7	[11:aa	aw	ne	oe

	A5, DUR. 7/1	A6, DUR. 1/5	A7, DUR. 2/3	A8, DUR. 4.5/6	A9, DUR 3.5/4
S1		aw)ii ae	oe aa ne]			
S2	oe]	[RAO20:oo	uu aw ii ae	oe	aa	ee	ne]
S3	aa uu (aw	ne	oo]		
S4	ae	ee)]			
S5	[SOo2:ne	aa		oo ee aw ae	uu ii oe]
S6| ii (ee	oe		ne	uu |oo aa aw]
S7	oo	uu	ee	ii	ae]

in retrograde). The actual durations change from variation to variation and from the first to the second half within single variations.

The first aggregate is distinctive in that seven colors of the prime order of the series are presented consecutively in a single linear voice. Since this matrix is used in reverse order for the second half of the basic structure, the retrograde of that same sequence ends the basic structure—once more in a single voice. The beginning and ending of each variation are similar and distinctive as a result of this emphasis on "color melody" derived from the structure of the first aggregate.

Another consequence of the overall form of the basic structure is the juxtaposition of aggregate nine and its retrograde at the midpoint of the structure. In many of the variations the [ae] color in the seventh strand punctuates the exact center of the structure—the single, prominent, usually low-pitched "note" serving both the direct and retrograde aggregates.

A pair of features of the structure that serve as landmarks are aggregate six in the direct half of the structure and aggregate five in the retrograde half. Here the only exceptions to strictness of the retrograde are swapping of [ee] and [aw] between aggregates five and six in the second half (indicated by the parentheses in Table 7). These two aggregates—the only ones with at least one representative from each of the seven linear strands—are the "thickest" vertically and, at the same time, the most active in the structure because they are the briefest in relative duration.

Setting the Combinatorial Color Structure

The combinatorial matrix of Table 7 provides the sound-color structure for *Colors*, just as similar matrices form the pitch and/or the rhythmic structure in other works with combinatorial structures. The realization or "setting" of the serial structure, whether it is sound color or pitch that is being serially controlled, consists of choosing values for the remaining "free" musical elements. For some composers, in whose music the serial structure is made to encompass effectively all the musical elements,[14] the final realization of the structure is trivial. In *Colors*, on the other hand, certain musical parameters—loudness, articulation, source-spectrum characteristics and, to some extent, duration and pitch—are purposely left out of the precompositional design so that those parameters can be used to emphasize, clarify, and otherwise highlight aspects of the sound-color structure.

The groups of variations express the sound-color structure in strikingly different ways. To illustrate certain ways that sound color can be coordinated with other musical parameters, the first group of three variations is discussed below in some detail; the remaining groups are treated more briefly.

[14] Certain of Babbitt's works and Boulez's *Structures* approach this kind of total serialization.

The Settings in Variations 1–3

Figure 41 is a score representing the sound colors, pitches, and rhythms of the first three variations. The eighteen aggregates of the combinatorial matrix and its retrograde are separated by the heavy vertical lines, with the accumulative time in arbitrary temporal units indicated at the top of each line. The durations of the temporal units—fixed over each half of the structure, but differing from variation to variation—are given in Table 8.

RECORDED
EXAMPLE 9

The rhythm of the basic structure is constrained by a rule imposed on the selection of durations within each linearly presented series of the nine-color set. This rule requires that each of the first seven multiples of a fixed subunit of duration be assigned to seven of the nine colors, the remaining two colors having free durations. Although fixed in each voice, the durations of the subunits increase from voice one to voice seven: the briefest durations occur in voice one, somewhat longer

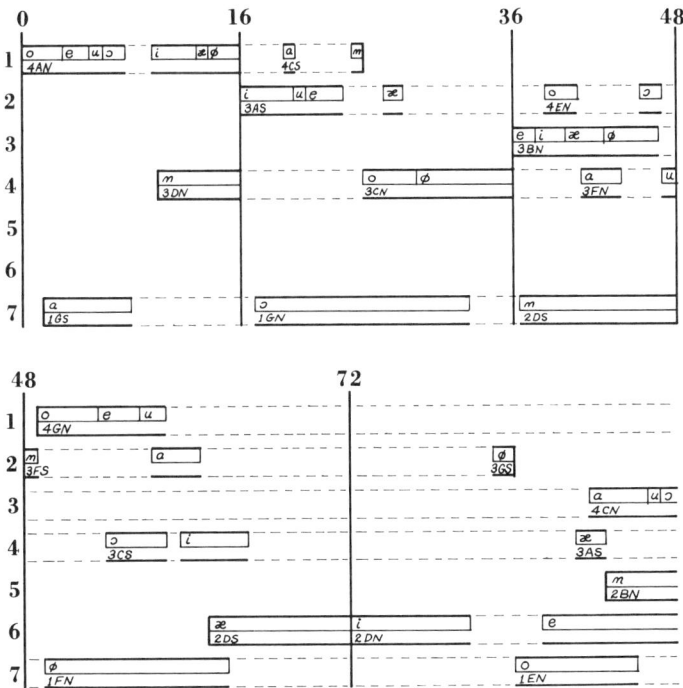

FIGURE 41: *Score of Variations 1–3 of Slawson's* Colors.

218   SOUND COLOR

FIGURE 41: (Continued)

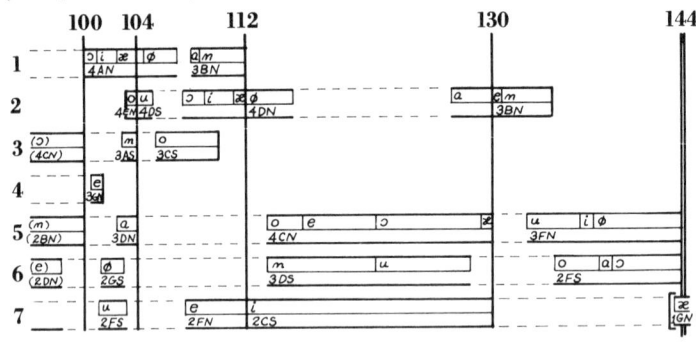
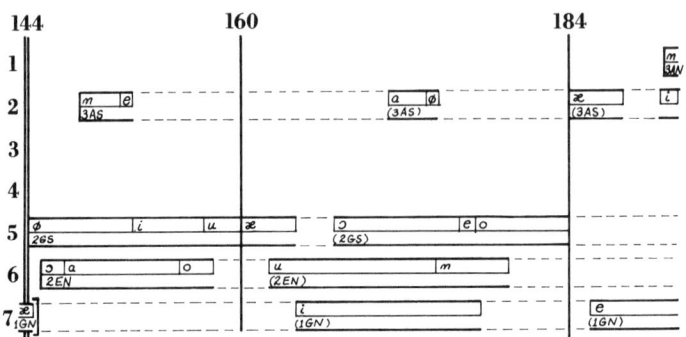
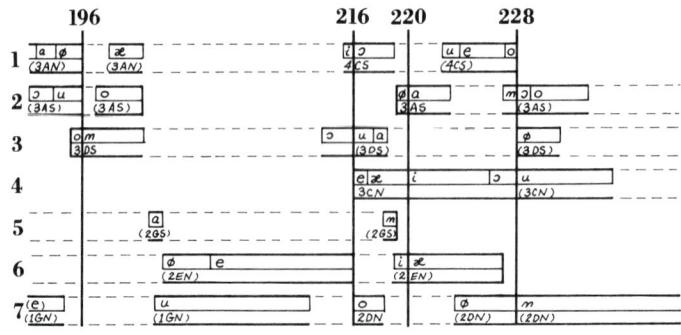
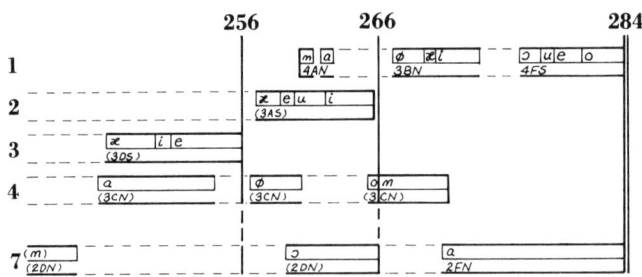

TABLE 8: *Durations of the units in the first section of* Colors

	VAR. 1	VAR. 2	VAR. 3
First half	0.54	0.48	0.42
Second half	0.24	0.30	0.36

durations in voice two, etc., with the longest durations in voice seven. The order of the seven fixed durations and the two free durations are not restricted in the compositional design—the order, in other words, is "free." Events can thus be aligned among the voices during the final, "realization" phase of the composition in a traditional way—e.g., punctuating gestures, sustaining particular color mixtures, or emphasizing or obscuring aggregate boundaries.

The pitches, in the first half of the combinatorial structure in these variations, are kept the same in each voice throughout a single vertical aggregate of sound colors. At the aggregate boundaries the pitches in all the voices are changed. Thus, each aggregate is characterized by a "chord": Aggregate 1 by G♯, D♮ and A♮; Aggregate 2 by G♮, C♮, A♯, and C♯; and so on. In the second half of the variations, a single pitch is assigned to an entire linear presentation of the color series.[15] For example, at temporal unit 144 and following in voice two, the A♯ is carried through until the end of the series at approximately temporal unit 200. This scheme for the pitch structure emphasizes the vertical aggregates in the first halves of these variations and emphasizes the linear series in the second halves, a feature that strongly differentiates the first from the second half of the combinatorial structure.

Certain rhythmic "motives" that stand out in the contrapuntal structure point up subsets of the color series that are transformations of each other. The rhythmic setting of the [ii ae oe] in voice one at the end of the first aggregate is repeated, with only slightly longer dura-

---

[15]The single exception to this rule is in the slowest seventh voice, in which a total of three pitches were assigned to the single series distributed over the entire half-structure.

tions, for the [ii uu ee] of voice two immediately following in the second aggregate. These subsets, at different order positions in their respective color series, are SMALLNESS inversions of each other. The next occurrence of the same "long-short-long" rhythmic pattern is at the end of Aggregate 5 in voice three. The colors in this case, [aa uu aw], are the retrograde of the IO(IA) of the three-element set of the same rhythm in Aggregate 1.

As is indicated in Table 4, the source character in each of these three variations is the same over half of the combinatorial structure. In Variation 1, the voices are initially entirely pitched; then noise is introduced softly and gradually. At the halfway point, the relation between noise and pitched sources is reversed; noise is prominent and the pitched sources are in the background. The pitched sounds of Variation 1 are transformed in Variation 2 by rather fast amplitude modulation and, in some cases, by glissando-like frequency modulation. The pitch can be heard clearly, for the most part, but it is not as prominent as it was in Variation 1. In Variation 2, noise is in the background in the first half and in the foreground in the second half. The relationship of pitch and noise in Variation 1 is reversed in Variation 3: noise dominates the first half of the variation; sustained pitched sources, the second. Because pitch can be used to differentiate voices (by assigning, say, one pitch to a voice) and noise cannot, it follows that the "linear" presentation of the second half of the combinatorial matrix, excited for the first time by a pitched source, is emphasized strongly, also for the first time, in Variation 3. The source characteristics thus provide both contrast among the first three variations and unification of the three into a balanced group.

RECORDED
EXAMPLE 10

The Remaining Variations: A Summary

The first three variations, with their steady-state sound colors, present and emphasize the sound-color structure repeatedly. This emphasis of the structure permits a change to dynamic or glissando colors in Variations 4 through 7. In these variations, the same color structure with the same relative durations of the aggregates is preserved. Here the colors of the combinatorial matrix become the endpoints of continuous changes in the filter resonances. The sources are also continuous, with approximately equally spaced durations assigned to each color within each aggregate.

Although less clear in Variations 4 through 7 than in Variation 1–3, the color structure can be followed in this second group. In Variation 4, staccato chords quietly punctuate each aggregate. In Variations 5 and 6, the color structure is detectable largely by the contrasting temporal density of the color aggregates. The brief, "thick" Aggregates 6 and 14, with their diphthong-like colors, are the most prominent characteristics. Variation 7, repeating the prestissimo of Variation 6 but with pitched sources, paradoxically results in an emphasis on the pitch structure. The sound colors in this variation, changing at speeds comparable to formant transitions in consonants or to attack transients in musical instruments, recede to the background.

The pulsed sources of Variations 8–10 expose another means of excitation for the filters that carry the sound-color structure. The pulses are synthesized by switching signals from periodic oscillators briefly on and off. The pulses, therefore, have a pitch. The sound colors of most of the texture in all three of these variations are continuously changing, as in Variations 4–7. The remaining strands, made up of steady-state colors, are drawn from different combinations of voices in each of the variations in this group. The steady-state strand in Variation 8 is largely from voices two, three, and four of the combinatorial matrix; it is mostly from voices five, six, and seven in Variation 9; and it is from the most active portions of voices one, two, and three in Variation 10. The source in the steady-state strand is always noise. The effect of this increasingly prominent strand is to reintroduce the steady-state sound colors of the first group of variations.

The final variation returns to completely steady-state sound colors. The sources in this variation are a mixture of pitched sound and noise throughout. Reversing the pitch settings of Variations 1–3, however, the first half of the combinatorial structure of Variation 11 is set linearly, with single pitches following the horizontal sound-color series across aggregate boundaries.[16] The second half of the structure is

RECORDED
EXAMPLE 11

[16]The seventh line is an exception: the pitches of that line change twice during the first half of the color structure, as they did in the second half of the structure in Variations 1–3.

set vertically, the pitches changing with each aggregate. The return to essentially the same presentation of the color structure that was heard repeatedly in the first group of variations is accomplished in a fresh way in Variation 11 by reversing the function of pitch in the two halves of the variation.

CONCLUSION AND PERSPECTIVES

Sound color appears to be a compositionally viable musical element that can be subjected to precise structural control in ways that are analogous to those developed for pitch. Certain sound-color collections, having specific cardinalities and geometric relationships, provide for a wide range of formally well-behaved transformations. In those collections, a version of combinatoriality can be applied to sound color. Some operations, such as transposition with respect to LAXNESS and "stretching" of the sound-color space, appear to be independent of the other operations and can be employed extrinsically to the regular sound-color structure. Musical application of some of these "set-theoretic" possibilities, as demonstrated in the author's *Colors*, appears to hold promise.

The most interesting musical questions posed by the theory of sound color and the investigation of the operations reported in this chapter revolve around the interaction of sound colors with the various characteristics of the source—most significantly, pitch. In *Colors*, the usual relationship between pitch and timbre in instrumental music is largely reversed; changes in pitch are relatively few, whereas the sound color changes relatively rapidly. However, the interaction of the two elements is quite straightforward. *Colors* can be construed as a direct descendant of Schoenberg's "Farben" movement from Opus 16. In that work, too, "color"—that is to say, instrumental combinations—is rather simply related to pitch (Burkhart 1973). Clearly, the relationships between color and pitch need not be so simple. It will be interesting to see whether composers can work out intricate and elaborate reciprocal relationships between sound color and pitch that are effective in building musical structures.

In poetry, the sonic emphasis is on the phonetic qualities of the words, which are largely determined by the characteristics of the vocal filter. In music, the emphasis is on pitch, a property of the source. There have been meetings between the two—song, text-sound compositions—but few in which integration is intricate and in which speech sounds and pitches are equal in musical importance. Words are set to music, but the sounds of the words, typically, are taken by the composer as a precondition of the musical composition. The exceptions to this general rule—Babbitt's *Phonemena* and, possibly, vocal settings of poetry in tonal languages—suggest some of the possibilities of a more intimate meshing of speech sounds and pitch. One barrier to integration may derive from the semantic aspects of poetry; the sounds are from words, phrases, and sentences that have meaning.

The purely structural use of speech sounds—in which, for example, musical operations are applied to them—is another matter. The vowels in a line of poetry could be treated as a series of sound colors and further lines could then be restricted to transformations of the original vowel series that are specified by the theory (Rules 3a and 3b). Poets may find that their interest in allusion is too greatly restricted by use of such "musical" operations. Another equally important factor has impeded musicians' exploration of this no-man's-land. There has been no well-developed, perceptually compelling canon of laws that could form the basis of abstract musical processes involving speech sounds. The present study can be construed as a beginning of that codification, but only a beginning—one that leaves many questions unanswered and many interesting creative possibilities unpursued.

CHAPTER EIGHT

A Critique and Afterword

THE THEORY OF SOUND COLOR as developed in this study is unfinished. It is purposely vague where too little is known to be specific, but the vagueness is nevertheless a fault. The theory may also simply be wrong in some respects. For example, SMALLNESS may not have sufficient perceptual salience to make it as strong an attribute of color as the other dimensions, and the theory may have wrongly confounded SMALLNESS with "stretching" the sound-color space. The music-theoretical advantages of retaining SMALLNESS have been made clear, however. Transposition is a problematic operation on sound color. The wrap-around feature of that operation is necessary theoretically, but it is perceptually counterintuitive.

Although the dimension of LAXNESS is somewhat problematic, its definition as an aspect of sound color is an important contribution of this study. Another is the discovery that inversions of sound colors appear to be powerful carriers of musical invariance. But the empirical verification of these aspects of the theory, from scientific experimentation or from application in music, remains incomplete.

The more fundamental part of the theory—the part that defines sound color psychoacoustically and claims its independence from other auditory and musical parameters—is quite strongly supported.

Physiology, psychoacoustics, speech science, and musical practice all provide evidence of the separate existence of sound color as an attribute of sound. Not all the postulated dimensions of sound color are as well supported, and for the operations on sound color, only slight hints of confirmation can be found in the scientific literature and the musical repertory.

The theory of sound color, on the other hand, is unlike most music theories in which the aim is to understand existing musical artifacts. It has more in common with scientific theories in which a mathematical/logical derivation leads to predictions about experiments that have not yet been performed. Many necessary experiments with sound color—both scientific and musical—remain to be carried out. This study will have served its purpose if it stimulates scientists and musicians to perform some of those experiments.

APPENDIX

Using Filters to Control Sound Color in Electronic and Computer Music

MUSIC PRODUCED BY ELECTRONIC MEANS—often called electroacoustic music—is basically of two types: electronic music made with voltage-controlled, analog synthesizers, and computer-synthesized music. The simulation and control of filters in these two cases is similar in principle but different in detail. The purpose of this appendix is to introduce the use of filters in both voltage-controlled synthesizers and digital computers in a practical, non-mathematical way. (A standard text presenting a more thorough and mathematically sophisticated treatment of digital filters is Rabiner and Gold 1975.)

FILTERS IN THE ANALOG ELECTRONIC MUSIC STUDIO

Filters in the analog electronic music studio are modules with one or more inputs and one or more outputs. The best filter module to control sound color is the kind in which the center or cutoff frequency of the filter is controlled by an exterior voltage—so-called voltage-controlled filters (VCFs). At least two such filters are required and as many as four or five can be put to good use. These filters usually provide control over bandwidth (sometimes called Q or resonance in VCFs) and they often give the user a choice of using the filter in low-pass, band-pass, or high-pass mode.

To control sound color, the source is connected to a network of these filters connected in series, taking the low-pass output of each filter (see Figure 42). This connection of a finite number of low-pass filters will result in an abnormal attenuation of the frequencies of the source that are higher than the highest frequency filter (see Chapter Two). To compensate for this effect, a band-reject ("notch") filter can be set at a frequency above that of the highest low-pass filter.[1] This has the effect of a "higher-pole correction" circuit (Rosen 1960). Although in theory it makes no difference in what order the filters are connected, better signal-to-noise ratios are obtained if the filters are ordered with the higher frequencies closest to the source. In practice the sound color can be enhanced artificially by mixing the band-pass outputs of each filter with the output of the low-pass filter network. This use of parallel filtering will introduce antiresonances into the spectrum envelope which also tend to correct the high-frequency attenuation of the network.

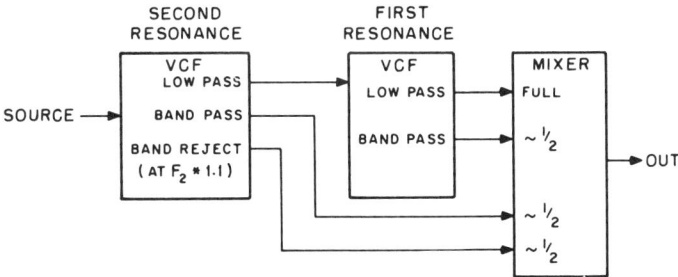

FIGURE 42: *Filter network in an analog electronic music studio.*

As a general rule the bandwidths of the filters do not have a large effect on sound color except at the extremes. Too wide bandwidths result in rather dull sound colors that are hard to differentiate; too narrow bandwidths result in ringing and/or distortion when components of the source correspond to the frequencies of the filters. As a general rule, the bandwidths of the filters should be set to about 10 percent of the cutoff frequency.

[1] Alternatively, the band-reject filter can be set at a very low frequency with its output passed through the F2 filter. This approximates another way of correcting the slope of the spectrum envelope: the "zero" at zero frequency.

Control of the filter network is problematic in an entirely analog studio. It can be managed with some difficulty by using a sequencer that supplies at least two time-locked control signals, one for each of the first two formant filters. A more flexible setup is available in a hybrid computer-controlled studio. In such a studio the signals that control the cutoff frequencies of the filters (and the bandwidths, if desired) are generated by the computer digitally and transformed by means of digital-to-analog converters into analog control voltages. The Computer and Electronic Music Studio at the University of Pittsburgh (in which the piece *Colors* was realized) provides for computer control of analog synthesizers in this manner.

FILTERS IN COMPUTER MUSIC

A great deal of work is currently being done on the use of digital filters in electronic music (e.g., Smith and Angel 1982). The following represents only a few hints about ways to design resonance filters of the type used in vowel synthesis for speech.

The basic problem is to calculate the amplitude of a series of samples of the wave form that results from an arbitrary source exciting a filter network. The calculations are carried out by writing a computer program that solves a "difference equation"—roughly, the digital form of the differential equation that describes the effect of a single resonance filter:

$$x_t = Ax_{t-\Delta t} + Bx_{t-2\Delta t} + F_{t-\Delta t}$$

where x_t is the value of the sample that is t samples from the beginning of the sound being synthesized, Δt is the time in seconds between samples, F_t is the value of the source at sample t, and A and B are coefficients that determine the frequency and bandwidth of the resonance. These coefficients are:

$$A = 2e^{-2\pi b \Delta t} \cos 2\pi f \Delta t$$
$$B = -e^{-4\pi b \Delta t}$$

where b is the bandwidth in Hz and f is the center frequency in Hz of the resonance.

The computer program begins by setting $x_{t-\Delta t}$, $x_{t-2\Delta t}$, and $F_{t-\Delta t}$ to zero, calculating x_t (which will, of course, be zero), and storing that value in an output table. Then $x_{t-\Delta t}$ is set to the value of x_t, $F_{t-\Delta t}$ is set to the value of the first sample of the source, and a new value of x_t is calculated and stored. Then $x_{t-2\Delta t}$ is set to the value of $x_{t-\Delta t}$, $x_{t-\Delta t}$ is set to the value of x_t, $F_{t-\Delta t}$ is set to the value of the second sample of the source, and a third value of x_t is calculated and stored. This cycle of operations continues until the end of the segment of music or until it is interrupted by the need to change some parameter, such as A or B—controlling the filter frequency and bandwidth, or F—the frequency, amplitude, etc., of the source.

The algorithm just described will simulate the effects of a single resonance filter. To synthesize sounds with multiple resonances (connected in series), the output of one filter stage (i.e., the X_t's) becomes the source for a second filter; the output of the second filter becomes the source for the third filter, and so on. There is no limit to the number of resonances that can be calculated in this manner except the practical one: each resonance requires at least two multiplications and two additions for each sample. At a sampling rate of, say, 40,000 samples per second, then, a single musical voice in which the source is provided and a four-resonance filter is used will require 320,000 multiplications and 320,000 additions for each second of sound.

Several programs are available that implement this algorithm (or one similar to it) in a form that makes it possible for the composer without programming experience to specify filters—usually by specifying center frequencies and bandwidths. An early effort in this direction is described in Slawson (1969). Some researchers have also supplied "unit generators" (for programs based on the MUSIC system [Mathews 1969]) that simulate the effects of resonances. The RESON statement in the Music-11 program (designed by Barry Vercoe and available from Digital Music Systems, Boston, Massachusetts) is an example.

Some interest has been shown recently in implementing filter-like elements as hardware in special-purpose "signal processing" computers. In a system that contains such hardware filters, a separate computer is required to supply control instructions to the signal processor.

The process of control in such a system becomes similar to that in the computer-controlled analog studio discussed above.

Because there is much activity in the field of digital filtering as applied to music, it would be well for composers interested in implementing a digital-filter facility, for themselves or in their institution, to consult current issues of the *Computer Music Journal* or the *Journal of the Audio Engineering Society*.

Bibliography

American Standards Association, Inc. 1960. American Standard Acoustical Terminology, S1.1–1960.

Attneave, F., and R. K. Olson. 1971. Pitch as a Medium: A New Approach to Psychophysical Scaling. *American Journal of Psychology* **84**, 147–166.

Babbitt, M. 1955. Some Aspects of Twelve-Tone Composition. *The Score* **12**, 53–61.

———. 1960. Twelve-Tone Invariants as Compositional Determinants. *Musical Quarterly* **46**, 246–259.

———. 1962. Twelve-Tone Rhythmic Structure and the Electronic Medium. *Perspectives of New Music* **1**, 49–79.

———. 1964. *Composition for Synthesizer*. Columbia Princeton Electronic Music Center, Columbia MS–6566. (Sound recording.)

———. 1976. *Ensembles for Synthesizer*. Columbia Princeton Electronic Music Center, Finnadar QD–9010. (Quadraphonic sound recording, playable on stereo systems.)

———. 1977. "Phonemena." On *New Music for Virtuosos*, New World Records NW–209. (Sound recording.)

Barnwell, Y. M., E. M. Harris, B. J. Reagon, and Y. B. Williams. 1982. *Good News*. Flying Fish FF–245. (Sound recording. The group name of these singers is "Sweet Honey in the Rock.")

Baru, A. V. 1975. Discrimination of Synthesized Vowels [a] and [i] with Varying Parameters in Dog. Pp. 91–101 in G. Fant and M. A. A. Tatham, eds., *Auditory Analysis and Perception of Speech*. London: Academic Press.

Beach, D. 1969. A Schenker Bibliography. *Journal of Music Theory* **13**, 2–37.

———. 1979. A Schenker Bibliography: 1969–1979. *Journal of Music Theory* **23**, 275–286.

Békésy, G. von. 1957. Neural Volleys and Similarity between Some Sensations Produced by Tones and by Skin Vibrations. *Journal of the Acoustical Society of America* **29**, 1059–1069.

———. 1960. *Experiments in Hearing*. New York: McGraw-Hill.

Bengtsson, I. 1969. Review of *Traité des objets musicaux* by Pierre Schaeffer. *Svensk Tidskrift for Musikforskning* **51**, 177–181.

Bever, T. G., and R. J. Chiarello. 1974. Cerebral Dominance in Musicians and Nonmusicians. *Science* **185**, 537–539.

Bismarck, G. von. 1974a. Timbre of Steady Sounds: A Factorial Investigation of Its Verbal Attributes. *Acustica* **30**, 146–159.

———. 1974b. Sharpness as an Attribute of the Timbre of Steady Sounds. *Acustica* **30**, 159–172.

Boer, E. de. 1956. On the Residue in Hearing. Ph.D. thesis, University of Amsterdam.

———. 1976. On the "Residue" and Auditory Pitch Perception. Pp. 479–583 in W. D. Keidel and W. D. Neff, eds., *Handbook of Sensory Physiology*, Vol. V/3. Heidelberg: Springer.

———. 1977. Pitch Theories Unified. Pp. 321–334 in E. F. Evans and J. P. Wilson, eds., *Psychophysics and Physiology of Hearing*. London: Academic Press.

Boretz, B. 1974. *Group Variations: Music for Computers, Electronic Sounds and Players*. CRI SD–300. (Sound recording.)

Boring, E. G. 1942. *Sensation and Perception in the History of Experimental Psychology*. New York: Appleton-Century.

Bradshaw, J. L., and N. C. Nettleton. 1981. The Nature of Hemispheric Specialization in Man. *Behavioral and Brain Sciences* **4**, 51–91.

Britt, R. 1976. Intracellular Study of Synaptic Events related to Phase-Locking Responses of Cat Cochlear Nucleus Cells to Low Frequency Tones. *Brain Research* **112**, 313–327.

Britt, R., and A. Starr. 1976. Synaptic Events and Discharge Patterns of Cochlear Nucleus Cells. I. Steady-Frequency Tone Bursts. *Journal of Neurophysiology* **39**, 162–178.

Bruce, V. G., and M. J. Morgan. 1975. Violations of Symmetry and Repetition in Visual Patterns. *Perception* **4**, 239–249.

Brugge, J. F., and M. M. Merzenich. 1973. Patterns of Activity of Single Neurons of the Auditory Cortex in Monkey. Pp. 745–772 in A. Møller, ed., *Basic Mechanisms in Hearing*. New York: Academic Press.

Bullock, T. H., ed. 1977. *Recognition of Complex Acoustic Signals*. Berlin: Abakon.

Burdick, C. K., and J. D. Miller. 1975. Speech Perception by the Chinchilla: Discrimination of Sustained /a/ and /i/. *Journal of the Acoustical Society of America* **58**, 415–427.

Burkhart, C. 1973. Schoenberg's "Farben": An Analysis of Op. 16, No. 3. *Perspectives of New Music* **12**, 141–172.

Cardozo, B. L., ed. 1972. *Hearing Theory*. Eindhoven: Instituut voor Perceptie Onderzoek.

Chiba, T., and M. Kajiyama. 1941. *The Vowel, Its Nature and Structure*. Tokyo: Tokyo-Kaiseikan.

Chistovich, L. A. 1971. Problems of Speech Perception. Pp. 83–93 in L. L. Hammerich, R. Jakobson, and E. Zwirner, eds., *Form and Substance*. Copenhagen: Akademisk Forlag.

Christovich, L. A., and V. V. Lublinskaya. 1979. The "Center of Gravity" Effect in Vowel Spectra and Critical Distance between the Formants: Psychoacoustical Study of the Perception of Vowel-like Stimuli. *Hearing Research* **1**, 185–195.

Chomsky, N., and M. Halle. 1968. *The Sound Pattern of English*. New York: Harper and Row.

Cogan, R. 1980. Tone Color: The New Understanding. *Sonus* **1**, 3–24.

Cogan, R., and P. Escot. 1976. *Sonic Design: The Nature of Sound and Music*. Englewood Cliffs, N.J.: Prentice-Hall.

Cole. R. A., A. I. Rudnicky, V. W. Zue, and D. R. Reddy. 1980. Speech as Patterns on Paper. Pp. 3–50 in R. A. Cole, ed., *Perception and Production of Fluent Speech*. Hillsdale, N.J.: Lawrence Erlbaum Associates.

Cooper, L. A., and P. Podgorny. 1976. Mental Transformations and Visual Comparison Processes: Effects of Complexity and Similarity. *Journal of Experimental Psychology: Human Perception and Performance* **2**, 503–514.

Cooper, L. A., and R. N. Shepard. 1975. Mental Transformations in the Identification of Right and Left Hands. *Journal of Experimental Psychology: Human Perception and Performance* **1**, 48–56.

Curry, F. K. W. 1967. A Comparison of Left-Handed and Right-Handed Subjects on Verbal and Non-Verbal Dichotic Listening Tasks. *Cortex* **3**, 343–352.

Cutting, J. E., and B. S. Rosner. 1974. Categories and Boundaries in Speech and Music. *Perception and Psychophysics* **16**, 564–570.

Cutting, J. E., B. S. Rosner, and C. F. Foard. 1976. Perceptual Categories for Musiclike Sounds: Implications for Theories of Speech Perception. *Quarterly Journal of Experimental Psychology* **28**, 361–378.

Darwin, C. J. 1969. Auditory Perception and Cerebral Dominance. Ph.D. thesis, University of Cambridge.

Deutsch, D. 1978. The Psychology of Music. Pp. 191–224 in E. C. Carterette and M. P. Friedman, eds., *Handbook of Perception*. New York: Academic Press.

Dodge, C. 1976a. *In Celebration: Synthesized Speech Music.* CRI SD–348. (Sound recording.)

———. 1976b. "Speech Songs." On *Synthesized Speech Music*, CRI SD–348. (Sound recording.)

———. 1976c. "The Story of Our Lives." On *Synthesized Speech Music*, CRI SD–348. (Sound recording.)

Druckman, J. 1971. *Synapse/Valentine.* Nonesuch H71253. (Sound recording.)

Ehresman, D., and D. L. Wessel. 1978. *Perception of Timbral Analogies.* IRCAM Technical Report 13.

Eldredge, D. H. 1974. Inner Ear–Cochlear Mechanics and Cochlear Potentials. Pp. 549–584 in W. D. Keidel and W. D. Neff, eds., *Handbook of Sensory Physiology*, Vol. V/1. Heidelberg: Springer.

Erickson, R. 1975. *Sound Structure in Music.* Berkeley: University of California Press.

Erulkar, S. D. 1975. Physiological Studies of the Inferior Colliculus and Medial Geniculate Complex. Pp. 145–198 in W. D. Keidel and W. D. Neff, eds., *Handbook of Sensory Physiology*, Vol. V/2. Heidelberg: Springer.

Evangelisti, F. 1967. Review of *Traité des objets musicaux* by Pierre Schaeffer. *World Music* **9**, 61–63.

Evans, E. F. 1975. Cochlear Nerve and Cochlear Nucleus. Pp. 1–108 in W. D. Keidel and W. D. Neff, eds., *Handbook of Sensory Physiology*, Vol. V/2. Heidelberg: Springer.

———. 1977. Frequency Selectivity at High Signal Levels of Single Units in Cochlear Nerve and Cochlear Nucleus. Pp. 185–196 in E. F. Evans and J. P. Wilson, eds., *Psychophysics and Physiology of Hearing*. London: Academic Press.

———. 1978. Place and Time Coding of Frequency in the Peripheral Auditory System: Some Physiological Pros and Cons. *Audiology* **17**, 369–420.

Evans, E. F., and J. P. Wilson, eds. 1977. *Psychophysics and Physiology of Hearing*. London: Academic Press.

Fant, G. 1959. Acoustic Analysis and Synthesis of Speech with Applications to Swedish. *Ericsson Technics* **1**, 3–108.

———. 1960. *Acoustic Theory of Speech Production*. 's-Gravenhage: Mouton.

———. 1967. Auditory Patterns of Speech. Pp. 111–125 in W. Wathen-Dunn, ed., *Models for the Perception of Speech and Visual Form*. Cambridge, Mass.: MIT Press.

———. 1973a. A Note on Vocal Tract Size Factors and Nonuniform F-Pattern Scalings. Pp. 84–93 in G. Fant, ed., *Speech Sounds and Features*. Cambridge, Mass.: MIT Press.

———. 1973b. The Nature of Distinctive Features. Pp. 151–159 in G. Fant, ed., *Speech Sounds and Features*. Cambridge, Mass.: MIT Press.

Fant, G., and M. A. A. Tatham, eds. 1975. *Auditory Analysis and Perception of Speech*. London: Academic Press.

Fennelly, B. 1967. A Descriptive Language for the Analysis of Electronic Music. *Perspectives of New Music* **6**, 79–95.

Flanagan, J. L. 1955. A Difference Limen for Vowel Formant

Frequency. *Journal of the Acoustical Society of America* **27**, 613–617.

———. 1965. *Speech Analysis, Synthesis and Perception.* New York: Academic Press.

Flanagan, J. L., and M. G. Saslow. 1958. Pitch Discrimination for Synthetic Vowels. *Journal of the Acoustical Society of America* **30**, 435–442.

Forte, A. 1964. A Theory of Set Complexes for Music. *Journal of Music Theory* **8**, 136–183.

———. 1973. *The Structure of Atonal Music.* New Haven: Yale University Press.

———. 1978. *The Harmonic Organization of "The Rite of Spring."* New Haven: Yale University Press.

Fransson, F. 1966. *The Source Spectrum of Double-Reed Woodwind Instruments.* Technical Report, Speech Transmission Laboratory, QPSR 4/1966, Royal Institute of Technology (KTH).

Glarean, H. 1965. *Dodecachordon.* (Translated by C. A. Miller from the 1547 printing.) N.p.: American Institute of Musicology.

Goldstein, J. L. 1978. Mechanisms of Signal Analysis and Pattern Perception in Periodicity Pitch. *Audiology* **17**, 421–445.

Goldstein, M. H., Jr., and M. Abeles. 1975. Single Unit Activity of the Auditory Cortex. Pp. 199–218 in W. D. Keidel and W. D. Neff, eds., *Handbook of Sensory Physiology*, Vol. V/1. Heidelberg: Springer.

Goldstein, M. H., Jr., F. deRibaupierre, and G. H. Yeni-Komshian. 1971. Cortical Coding of Periodicity Pitch. Pp. 299–305 in M. B. Sachs, ed., *The Physiology of the Auditory System.* Baltimore: National Educational Consultants.

Gordon, H. W. 1970. Hemispheric Asymmetries in the Perception of Musical Chords. *Cortex* **6**, 387–398.

———. 1983. Music and the Right Hemisphere. Pp. 65–86 in A. W. Young, ed., *Functions of the Right Cerebral Hemisphere.* London: Academic Press.

Green, D. M. 1976. *An Introduction to Hearing.* Hillsdale, N.Y.: Lawrence Erlbaum.

Greenberg, S. 1980. *Neural Temporal Coding of Pitch and Vowel Quality: Human Frequency-Following Response Studies of Com-*

plex Signals. Technical Report, UCLA Working Papers in Phonetics 52, December.

Grey, J. M. 1977. Multidimensional Scaling of Musical Timbres. *Journal of the Acoustical Society of America* **61**, 1270–1277.

———. 1978. Timbre Discrimination in Musical Patterns. *Journal of the Acoustical Society of America* **64**, 467–472.

Grey, J. M., and J. A. Moorer. 1977. Perceptual Evaluation of Synthesized Musical Instrument Tones. *Journal of the Acoustical Society of America* **62**, 454–462.

Haggard, M. P. 1969. Perception of Semi-Vowels and Laterals. *Journal of the Acoustical Society of America* **46**, 115. (Abstract.)

———. 1977. Mechanisms of Formant Frequency Discrimination. Pp. 499–507 in E. F. Evans and J. P. Wilson, eds., *Psychophysics and Physiology of Hearing.* London: Academic Press.

Hanson, G. 1963. A Factorial Investigation of Speech Sound Perception. *Scandinavian Journal of Psychology* **4**, 123–128.

Haugeland, J. 1981. The Nature and Plausibility of Cognitivism. Pp. 243–281 in J. Haugeland, ed., *Mind Design.* Montgomery, Vt.: Bradford.

Heinichen, J. D. 1969. *Der Generalbass in der Komposition.* (Facsimile reprint of the Dresden edition of 1728.) Hildesheim: Olms.

Helmholtz, H. L. F. 1954. *On the Sensations of Tone as a Physiological Basis for the Theory of Music.* (Reprint of the second English edition, translated by A. J. Ellis and published in 1877.) New York: Dover.

Hienz, R. D., M. B. Sachs, and J. M. Sinnott. 1981. Discrimination of Steady-State Vowels by Blackbirds and Pigeons. *Journal of the Acoustical Society of America* **70**, 699–706.

Houtsma, A. J. M., and J. L. Goldstein. 1972. The Central Origin of the Pitch of Complex Tones: Evidence from Musical Interval Recognition. *Journal of the Acoustical Society of America* **51**, 520–529.

Howe, H. S., Jr. 1975. *Electronic Music Synthesis.* New York: Norton.

———. 1977. *Third Study in Timbre.* Opus One 47. (Sound recording.)

———. 1978. Timbral Structures for Computer Music. Pp. 214–225 in C. Roads, ed., *Proceedings of the International Computer Music Conference*. Evanston, Ill.: Northwestern University Press.

———. 1980. *Improvisation on the Overtone Series*. Opus One 53. (Sound recording.)

Hubel, D. H., and T. N. Wiesel. 1962. Receptive Fields, Binocular Interaction and Functional Architecture in the Cat's Visual Cortex. *Journal of Physiology* **160**, 106–154.

Huggins, W. H. 1952. A Phrase Principle for Complex-Frequency Analysis and Its Implications in Auditory Theory. *Journal of the Acoustical Society of America* **24**, 582–589.

Jakobson, R., G. Fant, and M. Halle. 1951. *Preliminaries to Speech Analysis: The Distinctive Features and Their Correlates*. Cambridge, Mass.: MIT Press.

Jansson, E. V. 1966. *Analogies between Bowed String Instruments and the Human Voice: Source-Filter Models*. Technical Report, Speech Transmission Laboratory, QPSR 3/1966, Royal Institute of Technology (KTH).

———. 1973. An Investigation of a Violin by Laser Speckle Interferometry and Acoustical Measurements. *Acustica* **29**, 21.

———. 1976. Long-Time-Average Spectra Applied to Analysis of Music. Part III: A Simple Method for Surveyable Analysis of Complex Sound Sources by Means of a Reverberation Chamber. *Acustica* **34**, 275–280.

Kaas, J. H., R. J. Nelson, M. Sur, C. S. Lin, and M. M. Merzenich. 1979. Multiple Representations of the Body within the Primary Somatosensory Cortex of Primates. *Science* **204**, 521–523.

Kahn, J. I., and D. H. Foster. 1981. Visual Comparison of Rotated and Reflected Random-Dot Patterns as a Function of Their Positional Symmetry and Separation in the Field. *Quarterly Journal of Experimental Psychology* **33A**, 155–166.

Kakusho, O., K. Kato, and T. Kobayashi. 1968. Just Discriminable Change and Matching Range of Acoustic Parameters of Vowels. *Acustica* **20**, 46–54.

Kay, N. 1968. Review of *Traité des objets musicaux* and *La musique concrète* by Pierre Schaeffer. *Tempo* **84**, 29–31.

Keidel, W. D. 1974. Information Processing in the Higher Parts of the

Auditory Pathway. Pp. 216–226 in E. Zwicker and E. Terhardt, eds., *Facts and Models in Hearing*. New York: Springer.

Kiang, N. Y. S. 1968. A Survey of Recent Developments in the Study of Auditory Physiology. *Annals of Otology, Rhinology, and Laryngology* **77**, 656–675.

Kiang, N. Y. S., T. Watenabe, E. C. Thomas, and L. F. Clark. 1965. *Discharge Patterns of Single Fibers in Cat's Auditory Nerve*. Cambridge, Mass.: MIT Press.

Kimura, D. 1964. Left-Right Differences in the Perception of Melodies. *Quarterly Journal of Experimental Psychology* **102**, 903–905.

———. 1967. Functional Asymmetry of the Brain in Dichotic Listening. *Cortex* **3**, 163–178.

Klein, W., R. Plomp, and L. C. W. Pols. 1970. Vowel Spectra, Vowel Spaces, and Vowel Identification. *Journal of the Acoustical Society of America* **48**, 999–1009.

Kruskal, J. B. 1964. Nonmetric Multidimensional Scaling: A Numerical Method. *Psychometrika* **29**, 115–129.

Kuhl, P. K. 1981. Discrimination of Speech by Nonhuman Animals: Basic Auditory Sensitivities Conducive to the Perception of Speech-Sound Categories. *Journal of the Acoustical Society of America* **70**, 340–349.

Ladefoged, P., and D. E. Broadbent. 1957. Information Conveyed by Vowels. *Journal of the Acoustical Society of America* **29**, 98–104.

Ladefoged, P., and R. Harshman. 1979. Formant Frequencies and Movements of the Tongue. Pp. 25–34 in B. Lindblom and S. Öhman, eds., *Frontiers of Speech Communication Research*. London: Academic Press.

Lane, Harlan L. 1965. The Motor Theory of Speech Perception: A Critical Review. *Psychological Review* **72**, 275–309.

Langner, G., D. Bonke, and H. Scheich. 1981. Selectivity of Auditory Neurons for Vowels and Consonants in the Forebrain of the Mynah Bird. Pp. 317–321 in J. Syka and L. Aitkin, eds., *Neuronal Mechanisms of Hearing*. New York: Plenum Press.

Lansky, P. 1982. *Six Fantasies on a Poem by Thomas Campion: Computer Music*. CRI SD-456. (Sound recording.)

Lansky, P., and K. Steiglitz. 1981. Synthesis of Timbral Families by Warped Linear Prediction. *Computer Music Journal* 5/3, 45–49.

Lass, R. 1976. *English Phonology and Phonological Theory.* Cambridge: Cambridge University Press.

Lauter, J. L. 1983. Stimulus Characteristics and Relative Ear Advantages: A New Look at Old Data. *Journal of the Acoustical Society of America* 74, 1–17.

Lehiste, I. 1972. The Units of Speech Perception. Pp. 187–235 in J. H. Gilbert, ed., *Speech and Cortical Functioning.* New York: Academic Press.

Lenneberg, E. H. 1962. Understanding Language without Ability to Speak: A Case Report. *Journal of Abnormal and Social Psychology* 65, 419–425.

———. 1967. *Biological Foundations of Language.* New York: Wiley.

Lewin, D. 1977. Forte's Interval Vector, My Interval Function, and Regener's Common-Note Function. *Journal of Music Theory* 21, 194–237.

Liberman, A. M., F. S. Cooper, D. P. Shankweiler, and M. G. Studdert-Kennedy. 1967. Perception of the Speech Code. *Psychological Review* 74, 431–461.

Liberman, A. M., P. C. Delattre, L. J. Gerstman, and F. S. Cooper. 1956. Tempo of Frequency Change as a Cue for Distinguishing Classes of Speech Sounds. *Journal of Experimental Psychology* 52, 127–137.

Liberman, A. M., K. S. Harris, H. S. Hoffman, and B. C. Griffith. 1957. The Discrimination of Speech Sounds within and across Phoneme Boundaries. *Journal of Experimental Psychology* 54, 358–368.

Liberman, A. M., and D. B. Pisoni. 1977. Evidence for a Special Speech-Perceiving Subsystem in the Human. Pp. 59–76 in T. H. Bullock, ed., *Recognition of Complex Acoustic Signals.* Berlin: Abakon.

Liberman, M. C. 1982. Single Neuron Labeling in the Cat Auditory Nerve. *Science* 216, 1239–1241.

Lichte, W. H. 1941. Attributes of Complex Tones. *Journal of Experimental Psychology* 28, 455–480.

Licklider, J. C. R. 1954. "Periodicity" Pitch and "Place" Pitch. *Journal of the Acoustical Society of America* **26**, 945. (Abstract.)

Lindblom, B., and S. Öhman, eds. 1979. *Frontiers of Speech Communication Research.* London: Academic Press.

Lindsey, P. H., and D. A. Norman. 1977. *Human Information Processing.* New York: Academic Press.

Luce, R. D. 1972. What Sort of Measurement Is Psychophysical Measurement? *American Psychologist* **27**, 96–106.

Luce, R. D., and E. Galanter. 1963. Discrimination. Pp. 191–243 in R. D. Luce, R. R. Bush, and E. Galanter, eds., *Handbook of Mathematical Psychology* I. New York: Wiley.

Manley, J. A., and P. Mueller-Preuss. 1981. A Comparison of the Responses Evoked by Artificial Stimuli and Vocalizations in the Inferior Colliculus of Squirrel Monkeys. Pp. 307–310 in J. Syka and L. Aitkin, eds., *Neuronal Mechanisms of Hearing.* New York: Plenum Press.

Marks, L. E., 1974. *Sensory Processes.* New York: Academic Press.

———. 1978. *The Unity of the Senses: Interrelations among the Modalities.* New York: Academic Press.

Marler, P. R. 1977. The Structure of Animal Communication Sounds. Pp. 17–36 in T. H. Bullock, ed., *Recognition of Complex Acoustic Signals.* Berlin: Abakon.

Marmor, G. S., and L. A. Zaback. 1976. Mental Rotation by the Blind: Does Mental Rotation Depend on Visual Imagery? *Journal of Experimental Psychology: Human Perception and Performance* **2**, 515–521.

Mathews, M. V. 1969. *The Technology of Computer Music.* Cambridge, Mass.: MIT Press.

Merzenich, M. M., and J. H. Kaas. 1980. Principles of Organization of Sensory-Perceptual Systems in Mammals. *Progress in Psychobiology and Physiological Psychology* **9**, 1–42.

Merzenich, M. M., G. L. Roth, R. A. Anderson, P. L. Knight, and S. A. Colwell. 1977. Some Basic Features of Organization of the Central Auditory Nervous System. Pp. 485–497 in E. F. Evans and J. P. Wilson, eds., *Psychophysics and Physiology of Hearing.* London: Academic Press.

Miller, J. D. 1977. Perception of Speech Sounds in Animals: Evidence for Speech Processing by Mammalian Auditory Mechanisms. Pp. 49–58 in T. H. Bullock, ed., *Recognition of Complex Acoustic Signals*. Berlin: Abakon.

Miller, J. D., and P. K. Kuhl. 1976. Speech Perception by the Chinchilla: A Progress Report on Syllable-Initial Voiced-Plosive Consonants. *Journal of the Acoustical Society of America* **59**, 554. (Abstract.)

Miller, J. R., and E. C. Carterette. 1975. Perceptual Space for Musical Structures. *Journal of the Acoustical Society of America* **58**, 711–720.

Miller, M. I., and M. B. Sachs. 1983. Representation of Stop Consonants in the Discharge Patterns of Auditory-Nerve Fibers. *Journal of the Acoustical Society of America* **74**, 502–517.

Miller, R. L. 1953. Auditory Tests with Synthetic Vowels. *Journal of the Acoustical Society of America* **25**, 114–121.

Møller, A. 1970. Two Different Types of Frequency Selective Neurons in the Cochlear Nucleus of the Rat. Pp. 168–174 in R. Plomp and G. F. Smoorenberg, eds., *Frequency Analysis and Periodicity Detection in Hearing*. Leiden: Sijthoff.

———. 1977. Coding of Time-Varying Sounds in the Cochlear Nucleus. *Audiology* **17**, 446–468.

Møller, A., ed. 1973. *Basic Mechanisms in Hearing*. New York: Academic Press.

Morley, T. 1953. *A Plain and Easy Introduction to Practical Music*. Edited by R. A. Harman from the 1597 edition. New York: Norton.

Morris, R. D. 1979–1980. A Similarity Index for Pitch-Class Sets. *Perspectives of New Music* **18**, 445–460.

Morse, P. M. 1948. *Vibration and Sound*. New York: McGraw-Hill.

Oliveros, P. 1968. *Sound Patterns: Extended Voices*. Odyssey 32160156. (Sound recording.)

Papcun, G., S. Krashen, D. Terbeek, R. Remington, and R. Harshman. 1974. Is the Left Hemisphere Specialized for Speech, Language, and/or Something Else? *Journal of the Acoustical Society of America* **55**, 319–327.

Peterson, G. E. 1952. Information-Bearing Elements of Speech. *Journal of the Acoustical Society of America* **24**, 629–637.

Peterson, G. E., and H. L. Barney. 1952. Control Methods Used in a Study of the Vowels. *Journal of the Acoustical Society of America* **24**, 175–184.

Pick, G. F. 1977. Comment on "Critical Bandwidth at High Intensities" by Scharf and Meiselman. Pp. 233–234 in E. F. Evans and J. P. Wilson, eds., *Psychophysics and Physiology of Hearing*. London: Academic Press.

Plomp, R. 1970. Timbre as a Multidimensional Attribute of Complex Tones. Pp. 397–414 in R. Plomp and G. F. Smoorenberg, eds., *Frequency Analysis and Periodicity Detection in Hearing*. Leiden: Sijthoff.

———. 1975. Auditory Analysis and Timbre Perception. Pp. 7–22 in G. Fant and M. A. A. Tatham, eds., *Auditory Analysis and Perception of Speech*. London: Academic Press.

———. 1976. *Aspects of Tone Sensation*. London: Academic Press.

Plomp, R., and G. F. Smoorenberg, eds. 1970. *Frequency Analysis and Periodicity Detection in Hearing*. Leiden: Sijthoff.

Plomp, R., and H. J. M. Steeneken. 1971. Pitch versus Timbre. Pp. 377–380 in *Proceedings of the Seventh International Congress on Acoustics, Budapest*. Budapest: Akademiai Kiadó.

Pols, L. C. W., H. R. C. Tromp, and R. Plomp. 1973. Frequency Analysis of Dutch Vowels from 50 Male Speakers. *Journal of the Acoustical Society of America* **53**, 1093–1101.

Rabiner, L. R., and B. Gold. 1975. *Theory and Application of Digital Signal Processing*. Englewood Cliffs, N.J.: Prentice-Hall.

Rameau, J.-P. 1971. *Treatise on Harmony*. Translated by P. Gossett from the 1722 Paris edition. New York: Dover.

Randall, J. K. 1967. Three Lectures to Scientists. *Perspectives of New Music* **5/2**, 124–140.

Risset, J.-C., and D. Wessel. 1982. Exploration of Timbre by Analysis and Synthesis. Pp. 25–58 in D. Deutsch, ed., *The Psychology of Music*. New York: Academic Press.

Ritsma, R. J. 1962. Existence Region of the Tonal Residue, I. *Journal of the Acoustical Society of America* **34**, 1224–1229.

———. 1963. Existence Region of the Tonal Residue, II. *Journal of the Acoustical Society of America* **35**, 1241–1245.

Ritsma, R. J., and A. Hoekstra. 1974. Frequency Selectivity and the Tonal Residue. Pp. 156–163 in E. Zwicker and E. Terhardt, eds., *Facts and Models in Hearing*. New York: Springer.

Robson, E. 1981. *Phonetic Music with Electronic Music*. Parker Ford, Pa.: Primary Press.

Rock, I. 1973. *Orientation and Form*. New York: Academic Press.

Rose, J. E., M. M. Gibson, L. M. Kitzes, and J. E. Hind. 1973. Studies of Phase-Locked Cochlear Output in Cells of the Anteroventral Nucleus in the Cochlear Complex of the Cat. Pp. 511–517 in A. Møller, ed., *Basic Mechanisms in Hearing*. New York: Academic Press.

Rosen, G. "Dynamic Analog Speech Synthesizer." 1960. Ph.D. thesis, Massachusetts Institute of Technology.

Rumelhart, D. E., and A. A. Abrahamson. 1973. Toward a Theory of Analogical Reasoning. *Cognitive Psychology* **5**, 1–28.

Sachs, M. B., and E. D. Young. 1980. Effects of Nonlinearities on Speech Encoding in the Auditory Nerve. *Journal of the Acoustical Society of America* **68**, 858–875.

Saldanha, E. L., and J. F. Corso. 1964. Timbre Cues and the Identification of Musical Instruments. *Journal of the Acoustical Society of America* **36**, 2021–2026.

Sanders, D. A. 1977. *Auditory Perception of Speech*. Englewood Cliffs, N.J.: Prentice-Hall.

Schaeffer, P. 1968. *Traité des objets musicaux*. Paris: Editions du Seuil.

Scharf, B. 1970. Critical Bands. Pp. 157–202 in J. V. Tobias, ed., *Foundations of Modern Auditory Theory*, Vol. 1. New York: Academic Press.

Schenker, H. 1979. *Free Composition*. Translated by Ernst Oster from the 1935 and 1956 editions. New York: Longman.

Schoenberg, A. 1912. *Five Pieces for Orchestra*, Opus 16. Composed in 1909, this work was revised in 1922. A second revision for smaller orchestra was completed in 1949. Frankfurt: Peters.

———. 1975. *Style and Idea*. Translated and edited by L. Stein. New York: St. Martin's Press.

———. 1978. *Theory of Harmony*. Translated by R. E. Carter from the 1911 edition. Berkeley: University of California Press.

Schouten, J. F. 1938. The Perception of Subjective Tones. *Proceedings of the Koninklijke Nederlandse Akademie van Wetenschappen* **41**, 1086–1093.

———. 1940. The Perception of Pitch. *Phillips Technical Review* **5**, 286–294.

———. 1970. The Residue Revisited. Pp. 41–58 in R. Plomp and G. F. Smoorenberg, eds., *Frequency Analysis and Periodicity Detection in Hearing*. Leiden: Sijthoff.

Searle, C. L. 1982. Speech Perception from an Auditory and Visual Viewpoint. *Canadian Journal of Psychology* **36**, 402–419.

Seebeck, A. 1841. Beobachtungen über einige Bedingungen der Entstehung von Tonen. *Annalen der Physik und Chemie* **53**, 417–436.

Shepard, R. N. 1966. Metric Structures in Ordinal Data. *Journal of Mathematical Psychology* **3**, 287–315.

———. 1974. Representations of Structure in Similarity Data: Problems and Prospects. *Psychometrika* **39**, 373–421.

———. 1980. Multidimensional Scaling, Tree-Fitting, and Clustering. *Science* **210**, 390–398.

———. 1982. Structural Representations of Musical Pitch. Pp. 334–390 in D. Deutsch, ed., *The Psychology of Music*. New York: Academic Press.

Shepard, R. N., and J. Metzler. 1971. Mental Rotation of Three-Dimensional Objects. *Science* **171**, 701–703.

Shower, E. G., and R. Biddulph. 1931. Differential Pitch Sensitivity of the Ear. *Journal of the Acoustical Society of America* **3**, 275–287.

Singh, S. 1976. *Distinctive Features: Theory and Validation*. Baltimore: University Park Press.

Singh, S., and D. R. Woods. 1971. Perceptual Structure of 12 American English Vowels. *Journal of the Acoustical Society of America* **49**, 1861–1866.

Slawson, A. W. 1968. Vowel Quality and Musical Timbre as Functions of Spectrum Envelope and Fundamental Frequency. *Journal of the Acoustical Society of America* **43**, 87–101.

―――. 1969. A Speech-Oriented Synthesizer of Computer Music. *Journal of Music Theory* **13**, 94–127.

―――. 1975. Sound, Electronics, and Hearing. Pp. 22–67 in J. H. Appleton and R. C. Perera, eds., *The Development and Practice of Electronic Music*. Englewood Cliffs, N.J.: Prentice-Hall.

―――. 1978. Review of *Sound Structure in Music* by Robert Erickson. *Journal of Music Theory* **22**, 105–109.

―――. 1981. The Color of Sound: A Study in Musical Timbre. *Music Theory Spectrum* **3**, 132–141.

―――. 1982. The Musical Control of Sound Color. *Canadian University Music Review* **3**, 67–79.

Smith, J. C., J. T. Marsh, S. Greenberg, and W. S. Brown. 1978. Human Auditory Frequency-Following Responses to a Missing Fundamental. *Science* **201**, 639–641.

Smith, J. O., and J. B. Angell. 1982. A Constant-Gain Digital Resonator Tuned by a Single Coefficient. *Computer Music Journal* **6/4**, 36–40.

Smoorenberg, G. F., and D. H. Linschoten. 1977. A Neurophysiological Study on Auditory Frequency Analysis of Complex Tones. Pp. 175–184 in E. F. Evans and J. P. Wilson, eds., *Psychophysics and Physiology of Hearing*. London: Academic Press.

Starr, D., and R. D. Morris. 1977–1978. A General Theory of Combinatoriality and the Aggregate, Parts 1, 2. *Perspectives of New Music* **16/1**, 3–35; **16/2**, 50–84.

Stevens, K. N. 1952. Frequency Discrimination for Damped Waves. *Journal of the Acoustical Society of America* **24**, 76–79.

―――. 1960. Toward a Model for Speech Recognition. *Journal of the Acoustical Society of America* **32**, 45–55.

―――. 1972. The Quantal Nature of Speech: Evidence from Articulatory-Acoustic Data. Pp. 51–66 in E. E. David and P. N. Denes, eds., *Human Communication: A Unified View*. New York: McGraw-Hill.

Stevens, K. N., and A. J. House. 1961. An Acoustic Theory of Vowel Production and Some of Its Implications. *Journal of Speech and Hearing Research* **4**, 303–320.

Stevens, S. S. 1951. Mathematics, Measurement, and Psychophysics.

Pp. 1–49 in S. S. Stevens, ed., *Handbook of Experimental Psychology*. New York: Wiley.

———. 1958. Problems and Methods in Psychophysics. *Psychological Bulletin* **55**, 177–196.

———. 1971. Issues in Psychophysical Measurement. *Psychological Review* **78**, 1577–1585.

Stevens, S. S., and H. Davis. 1938. *Hearing, Its Psychology and Physiology*. New York: Wiley.

Stevens, S. S., M. Guirao, and A. W. Slawson. 1965. Loudness, a Product of Volume times Density. *Journal of Experimental Psychology* **69**, 503–510.

Stevens, S. S., and J. Volkmann. 1940. The Relation of Pitch to Frequency: A Revised Scale. *American Journal of Psychology* **53**, 329–353.

Stevens, S. S., J. Volkmann, and E. B. Newman. 1937. A Scale for the Measurement of the Psychological Magnitude of Pitch. *Journal of the Acoustical Society of America* **8**, 185–190.

Stockhausen, K. 1962. The Concept of Unity in Electronic Music. *Perspectives of New Music* **1**, 39–48.

———. 1966. *Kontakte*. London: Universal.

———. 1963. *Kontakte*. DGG 138811. (Sound recording. Also available in a version with instruments: Wergo 60009.)

———. 1969a. *Hymnen*. DGG 2707039. (Sound recording, two-record set.)

———. 1969b. *Telemusik*. Vienna: Universal.

———. 1970. *Telemusik*. DGG 137012. (Sound recording.)

———. 1971. *Stimmung*. DGG 2543003. (Sound recording.)

Stravinsky, I. 1956. *Poetics of Music*. Translated by A. Knodel and I. Dahl. New York: Vintage.

Strong, W., and M. Clark. 1967. Synthesis of Wind Instrument Tones. *Journal of the Acoustical Society of America* **41**, 39–52.

Studdert-Kennedy, M. 1979. *The Beginnings of Speech*. Technical Report, Haskins Laboratory Status Report SR–58.

Studdert-Kennedy, M., A. M. Liberman, K. S. Harris, and F. S. Cooper. 1970. Motor Theory of Speech Perception: A Reply to Lane's Critical Review. *Psychological Review* **77**, 234–249.

Studdert-Kennedy, M., and D. Shankweiler. 1970. Hemispheric Specialization for Speech Perception. *Journal of the Acoustical Society of America* **48**, 579–594.

Stumpf, C. 1890. *Tonpsychologie*. Leipzig: Hirzel.

———. 1926. *Die Sprachlaute*. Berlin: Springer.

Subotnick, M. 1967. *Silver Apples of the Moon*. Nonesuch H71174. (Sound recording.)

———. 1976. *Until Spring*. Odyssey Y-34158. (Sound recording.)

———. 1980. *A Sky of Cloudless Sulphur*. Nonesuch H78001. (Sound recording.)

Suga, N., K. Kuzirai, and W. E. O'Neill. 1981. How Biosonar Information Is Represented in the Bat Cerebral Cortex. Pp. 197–219 in J. Syka and L. Aitkin, eds., *Neuronal Mechanisms of Hearing*. New York: Plenum Press.

Summerfield, A. Q., and M. P. Haggard. 1975. Vocal Tract Normalization as Demonstrated by Reaction Times. Pp. 115–141 in G. Fant and M. A. A. Tatham, eds., *Auditory Analysis and Perception of Speech*. London: Academic Press.

Sundberg, J., and E. Jansson. 1976. Long-Time-Average-Spectra Applied to Analysis of Music. Part II: An Analysis of Organ Stops. *Acustica* **34**, 269–274.

Swets, J. A. 1961. Is There a Sensory Threshold? *Science* **134**, 168–177.

Swift, R. 1975. Review of *Sound Structure in Music* by Robert Erickson. *Perspectives of New Music* **14**, 148–158.

Syka, J., and L. Aitkin, eds. 1981. *Neuronal Mechanisms of Hearing*. New York: Plenum Press.

Symmes, D. 1981. On the Use of Natural Stimuli in Neurophysiological Studies of Audition. *Hearing Research* **4**, 203–214.

Tenney, J. C., 1965. The Physical Correlates of Timbre. *Gravesaner Blatter* **26**, 106–109.

Terhardt, E. 1977. The Two-Component Theory of Musical Consonance. Pp. 381–390 in E. F. Evans and J. P. Wilson, eds., *Psychophysics and Physiology of Hearing*. London: Academic Press.

Thomas, I. B., P. B. Hill, F. S. Carroll, and B. Garcia. 1970. Temporal Order in the Perception of Vowels. *Journal of the Acoustical Society of America* **48**, 1010–1013.

Verbrugge, R. R., W. Strange, D. P. Shankweiler, and T. R. Edman. 1976. What Information Enables a Listener to Map a Talker's Vowel Space? *Journal of the Acoustical Society of America* **60**, 198–212.

Voigt, H. F., M. B. Sachs, and E. D. Young. 1981. Effects of Masking Noise on the Representation of Vowel Spectra in the Auditory Nerve. Pp. 113–118 in J. Syka and L. Aitkin, eds., *Neuronal Mechanisms of Hearing.* New York: Plenum Press.

———. 1982. Representation of Whispered Vowels in Discharge Patterns of Auditory-Nerve Fibers. *Hearing Research* **8**, 49–58.

Wessel, D. L. 1979. Timbre Space as a Musical Control Structure. *Computer Music Journal* **3/2**, 45–52.

Whitfield, I. C. 1967. *The Auditory Pathway.* London: Arnold.

———. 1980. The Relation between Pitch and Frequency in Complex Tones. Pp. 361–366 in G. van den Brink and F. A. Bilsen, eds., *Psychophysical, Physiological and Behavioral Studies in Hearing.* Delft, The Netherlands: Delft University Press.

Wuorinen, C. 1969. *Times Encomium.* Nonesuch H71253. (Sound recording.)

Yilmaz, H. 1967. A Theory of Speech Perception. *Bulletin of Mathematical Biophysics* **29**, 793–825.

Yost, W. A., and D. W. Nielson. 1977. *Fundamentals of Hearing.* New York: Holt, Rinehart, and Winston.

Young, E. D., and M. B. Sachs. 1979. Representation of Steady-State Vowels in the Temporal Aspects of the Discharge Patterns of Populations of Auditory-Nerve Fibers. *Journal of the Acoustical Society of America* **66**, 1381–1403.

Zwicker, E., and R. Feldtkeller. 1967. *Das Ohr als Nachrichtenempfänger.* Stuttgart: Hirzel.

Zwicker, E., and B. Scharf. 1965. A Model of Loudness Summation. *Psychological Review* **72**, 3–26.

Zwicker, E., and E. Terhardt, eds. 1974. *Facts and Models in Hearing.* New York: Springer.

Index

Abeles, M., 95
Abrahamson, A., 140
Absolute thresholds. *See* Thresholds
Acoustic measurements, limitations of, 179n4
Active theories of speech perception, 145–46
ACUTENESS: as a vowel feature, 52; contours of, 55–56; transposition in, 70–71; inversion in, 77–78; evidence for, 135–36, 158; and "sharpness," 137; and instrumental timbre, 139; musical variation in, 176–77
Aggregate, sound-color: in Stockhausen's *Telemusik*, 182; and pitch aggregates, 192; in *Colors*, 215–16; pitch punctuation of, 219. *See also* Twelve-tone Aggregates
Algorithm for solving difference equations, 229

Allure, 8, 67
Alternative axes of inversion, 82–83
Alternatives to sound color theory: based on psychoacoustics, 60; in vocal music, 185; in timbre pieces, 186–88
American Standards Association, 19
American vowels, acoustics of, 153
Amplitude of resonances, 33
Analog synthesizers. *See* Synthesizers
Analysis, speech-like, of non-speech sounds, 67
Analysis-by-synthesis theory of speech perception, 145–46
Analysis-synthesis method for studying timbre, 139
Anatomy of auditory system, 93–97
Angell, J., 228
Antiresonances, 42–43
Aphasia, 149
Articulation, and speech perception, 147

251

Artificiality of sound-color operations, 81
Artificial stimuli, effect on difference limens of vowels, 122
Attenuation patterns in simple resonators, 33
Attneave, F., 128
Attributes, psychological, 60
Auditory capacity as a prerequisite for speech, 62
Axes: of inversion, 77, 81–82; in visual symmetries, 161

Babbitt, M., 18, 213; *Composition for Synthesizer*, 169–70; *Ensembles for Synthesizer*, 177–80; *Phonemena*, 185
Band-pass filter, 34, 35n8, 168
Band-reject filter, 227
Bandwidth: of resonators, 32; in analog filters, 227; in digital filters, 228–29
Bark scale, 126
Barney, H., 53n25, 136, 153–54
Barnwell, Y., 185
Bar resonators, and sound-color dimensions, 66–67
Baru, A., 152
Basilar membrane, 94
Bassoon, resonances in, 157
Beach, D., 85n38
Behavior, hearing and, 144
Békésy, G., 94, 123–25
Bengtsson, I., 10
Bever, T., 150
Biddulph, R., 121n4
Biology: and phonetic universals, 60–61; and sound-color dimensions, 61; vs. learning, 120; as basis for language, 146
Bipartite structure of *Colors*, 214–16
Bismarck, G., 126, 136–38

Blend in sound-color mixtures, 89
Boer, E. de, 130
Bonke, D., 106–107, 112, 114
Boretz, B., *Group Variations*, 187–88
Boring, E., 47
Boulez, P., *Structures*, 216
Boundary conditions in uniform tubes, 35–36
Bounds of tube lengths, 43
Bradshaw, J., 151
Brain centers, functions in, 92
Brightness: as a dimension, 89, 134, 140; and "stretching" in vocal music, 186
Britt, R., 99, 110
Broadbent, D., 133
Bruce, V., 161
Brugge, J., 99
Buddhist chant, sound colors in, 181
Burdick, C., 151
Burkhart, C., 4n3, 222

Carterette, E., 138
Categorical perception: of speech sounds, 147–48; in animals, 152
Center frequency of a resonance, 32
"Center of gravity" effect, 125
Characteristic frequency of auditory neurons, 104–105
Chiarello, R., 150
Chiba, T., 35n9, 53n25
Children's vowels, and sound-color space, 155
Chistovich, L., 124–25
Chomsky, N., 51, 52n20, 53n24, 60–61
Chords, and brain laterality, 150
Chroma of a pitch, 128
Clarinet, 28–29

Clark, M., 31*n*5
Class of a musical object, 8
Clinical studies of hemispheric specialization, 148–49
Closure: and SMALLNESS transposition, 74; and musical completion, 180–83; in sound-color collections, 192–95; compositional implications, 198–99
Cochlea, frequency analysis in, 93–94
Cochlear nucleus, 95
Cochleotopic organization, 94, 95
Coding, of frequency in the auditory system, 109–111
Coefficients of difference equations, 228–29
Cogan, R., 11–13, 52*n*23, 89, 167–68
Cole, R., 167–68
Collections, color, 192
Color melody. *See* Melodies: color
Color mixture. *See* Mixture
Colors (Slawson), 207–222
Comb-filtered noise, cochlear nucleus response to, 104
Combinations of operations. *See* Operations: combinations of
Combinatoriality, sound color, 213–16
Commutativity of operations, 206
Compactness as a vowel feature, 52
Completion, musical, 182–83
Complexity of auditory system and physiological evidence, 91–92
Complex sounds, dimensions of, 134
Compositional theory and psychological theory, 84*n*37
Compound tube model, 39*n*12
Computer and Electronic Music Studio, 191

Computer control of analog filters. *See* Filters: in analog synthesis
Computer music: resultant sound color in, 187–88; filters in, 228–30
Computer programs for filter synthesis, 229–30. *See also* Filters
Computer Research in Music and Acoustics, Center for, 191*n*1
Confusions: among vowels, 135; between pitch and sound color, 171–72
Consonance, and sound-color mixture, 88
Consonants: sources in, 24*n*1; and sound-color dynamics, 87–88; auditory nerve responses to, 102*n*4; detectors for, 106; categorical perception in, 147–48; discrimination of, in animals, 152
Constancy, independent: vs. independent variation, 59–60; in musical treatment of dimensions, 176–77
Context: musical, and normalization, 66; in psychological experiments, 119; and discriminability, 123; and musical instrument timbre, 139
Continuity of sound-color dimensions, 52–53, 58
Contours. *See* Equal-value contours
Contralateral projections in auditory pathways, 95
Contrapuntal flexibility, 213
Contrapuntal lines in Babbitt's *Ensembles*, 180
Contrast in series of *Colors*, 210
Control, independent musical, of sound color, 168–75
Cooper, L., 163

Corroboration, musical, 175
Corso, J., 139
Cortex, auditory, 95; responses to vowels, 106–107; responses in bats, 107–108; hemispheric specialization in, 148–51
Corti's organ, 94
Counterpoint, of sound color, 169
Coupling: of source and filter, 28–31; in musical instruments, 30, 157
Criteria: for threshold measurements, 117–18; for music theories, 166–67
Critical band: and resonance DLs, 121; and "sharpness," 126
Cross-correlation in F-pattern detector model, 113–14
Cross-modality matching, method of, 118
Cross-section of tubes, 39
Curry, F., 150
Cutting, J., 148
Cyclic notation of operations, 192–95

Darwin, C., 150
Davis, H., 109, 128
DC amplitudes of resonances, 37
Decoupled resonance systems in musical instruments, 157
Detectability experiments, 118
Detectors, resonance, 112. *See also* Feature detectors in vision and hearing; F-pattern detectors
Deutsch, D., 84n37, 128
Dichotic listening, 149–51
Difference equations in computer synthesis, 228–29
Difference estimation of musical timbre and vowel quality, 130–31

Difference judgments, multidimensional, 125–26
Difference limen: measuring method, 117–18; for resonance frequency, 120–21; for resonance bandwidths, 121; for fundamental frequency, 121
Difference vs. invariance in music theory, 16–17
Differential thresholds. *See* Thresholds
Differentiation: of source from filter, 108–111; contrapuntal, by means of color, 168–70
Dimensions, sound color, 48–67; psychological vs. physical, definition of, 49; generality of, 53; as independent of psychoacoustics, 57; and operations, 68–69; possible additional, 89; physiological evidence for, 112; psychoacoustic evidence for, 134–140; musical evidence for, 175–83
Dimensions of musical instrument timbre, 138–40
Discriminable colors, number of, 122–23
Dissimilarity, judgments of, 135. *See also* Similarity
Dissonance, and sound-color mixture, 88
Distance relationships, and transposition, 70–74
Distinctive features: and sound-color dimensions, 51–54; and sound-color dynamics, 87–88
DL. *See* Difference limen
Dodge, C., 188
Domain: of theory, 18–20; of color-space shifts, 66

Dominant hemisphere, and lateralization, 149*n*2
Druckman, J., *Synapse*, 171–72
Drum, sound color in, 181–82
Durations, relation of to aggregates in *Colors*, 217–19
Dynamics, sound-color, 87–88

Echolocation in bats, 107
Edge of sound-color space: and wrap-around, 73; as axis of inversion, 83
Ehresman, D., 140
Eldridge, S., 94
Electronic music, 50, 165
Elements, musical, 15
Empirical methods, and dimensionality of timbre, 139–40
Empiricist tradition, 10
Encoding of resonances in cochlear nucleus, 98. *See also* Coding
English horn, resonances in, 157
Equal-ACUTENESS contours, 55–56
Equal-LAXNESS contours, 56
Equal-OPENNESS contours, 54–55
Equal-SMALLNESS contours, 56
Equal-value contours, 54–58; and transposition, 70; alternative for SMALLNESS, 74
Erickson, R., 10–11, 28
Error score, in analysis-by-synthesis theory, 145
Erulkar, S., 99, 110
Escot, P., 11–13, 89
Ethology, 61
Evangelisti, F., 5*n*5
Evans, E., 101, 104, 105, 111, 124
Evidence, compositional, status of, 190–91
Examples, for the source-filter model, 22–31

Excitation: in wall-tapping, 23; in the voice, 24–25; in musical instruments, 28. *See also* Source
Excitation pattern in mynah bird cortex, 107
Experience, and brain laterality, 150
Explanation in music theories, 166–67
Exposition, musical, of sound-color space, 177–80
Extended voices, technique of, and sound color, 185
Extensions of operations, 84–87

Facture in sound objects, 6–7
Fall-off, high-frequency: in multiple resonances, 37–38; and operation hierarchies, 85
Fant, G., 32*n*7, 35, 39*n*12, 40–41, 53, 63–64, 87, 153–54, 196
Feature detectors in vision and hearing, 99–100
Feature matching in speech perception, 146
Features: distinctive, 51–54; classificatory vs. phonetic, 53*n*24
Feldtkeller, R., 126
Fennelly, B., 67
FFR. *See* Frequency-following response
Filling, musical, of sound-color space, 178. *See also* Completion
Filters: and wall-tapping, 23; and source, 46–48, 130–33; in "natural" stimuli, 100; psychoacoustics of, 120–27; in electronic music, 165; in analog synthesis, 227; in computer music, 228–30
First-order neurons in the cochlea, 94

Fixed-frequency peaks in spectra of musical instruments, 140
Fixed-pitch theory of timbre, 47–48; evidence for, 132
Flanagan, J., 121, 122
Foard, C., 148
F1–F2 locus, 53, 153
Foreground, sound color in, 173–75
Formant transitions: categorical perception in, 147; in *Colors*, 221
Formants, 38–39; frequency of, 32; structure of, in musical instruments, 67; in auditory nerve response, 102; dichotic, and laterality, 150; reference frames in, as normalization cue, 156
Formant shift factor, and fundamental frequency, 131
Forte, A., 18, 129, 202n9
Foster, D., 161
Four-color subsets of normal eight-color collection, 206–207
F-pattern detectors, 106–108, 113–14
F-patterns, 39–41; as "natural" stimuli, 100
Fractionation, method of and the mel scale, 128
Fransson, F., 31n5, 140, 157
Free parameters, use of in *Colors*, 216
Frequency: spectrum, 25n2; mechanisms for discrimination of, 109–111; control of in difference equations, 228–29
Frequency analysis in the auditory periphery, 97
Frequency-following response, 111
Frequency-modulated source, use of in *Colors*, 208–209

F3. *See* Higher resonances: in vowels
Fullness, as dimension of complex sounds, 134
Fundamental frequency, 25n2, 26–27; in multidimensional scaling, 138; as a normalization cue, 156
Funneling, 124

Gagaku, sound colors in, 181
Galanter, E., 117, 118
Gamelan, sound colors in, 181
Genus of a musical object, 8
Glarean, H., 14n8
Glissandos of sound color, 88, 220–21
Gold, B., 226
Goldstein, J., 110, 111, 130
Goldstein, M., 95, 110
Gordon, H., 150, 151
Grain, 8, 11
Graphical analyses, 12–13, 167–68; Babbitt's *Ensembles*, 178–79
Grating acuity, 122n5
Gravity, 77
Green, D., 118
Greenberg, S., 111
Grey, J., 138–39
Group structure in composition, 198–99
Guidelines for source adequacy. *See* Source: adequacy of
Guirao, M., 60

Haggard, M., 121, 150, 156
Halle, M., 51, 52n20, 53n24, 60–61
Hanson, G., 158
HARDNESS, 89
Harmonics, pieces based on, 186–87
Harmonielehre (Schoenberg), 3–4

Harshman, R., 88*n*39
Haugeland, J., 92
Height, pitch, 128
Heinichen, J., 14*n*8
Helmholtz, H., 47*n*14
Hemispheric specialization of cortex, 148–51
Hienz, R., 152
Hierarchies, color, 85–87
"High encoding" in consonants, 148
Higher-pole correction, 38*n*11, 227
Higher resonances: in vowels, 53, 154; as cues for shifting, 63, 156; in auditory nerve response, 102–103
Hoekstra, A., 121*n*4
Holistic processes, and brain laterality, 150
House, A., 35*n*9, 38
Houtsma, A., 110, 130
Howe, H., *Third Study in Timbre*, 186–87
Hubel, D., 61*n*30, 99–100
Huggins, W. H., 43*n*13
Hybrid studios, control of filters in, 228

Ideal sources. *See* Source: ideal
Impulse response of resonator, 44–45
Impulse source. *See* Source: adequacy of
Independence of source. *See* Source: /filter model
Independent constancy. *See* Constancy
Independent control of sound color in music, 168–75
Inferior colliculae, 95–96, 99
Inflection, pitch, 24–25
Injury to brain, 149
Innate releasers, 61
Inner squares in sound-color collections, 197–98. *See also* Short vowels
Instrumental color, 19
Instrumental sources, musical treatment of, 181
Instrumentation, 4–5
Instruments, musical: and source/filter model, 27–31; dimensions of timbre in, 138; vowel-like resonances in, 157. *See also* Source
Integration, in cochlea, 124
Interdisciplinary research, 164
Internal representation in active theories of perception, 146
Interpretation of sound spectrographs. *See* Sound spectrographs
Interval: color, 77–78, 82; temporal, as neural response measure, 101
Intuition: in analysis, 167; in composition, 198–99
Invariance: vs. difference in music theory, 16–17; and operations, 18; of formants in auditory nerve, 102; in Rule 1, physiological evidence for, 109; sound-color, as contrapuntal differentiator, 169
Inversion, sound-color, 76–78, 79–80; and pitch inversion, 82, 160; musical evidence for, 183; cyclic definition of, 194–95; rules for combining, 201–202
Inversional invariance vector, 207
Ipsilateral projections in auditory pathways, 95
Irregular sources. *See* Source: adequacy of
Ives, C., 10

Jakobson, R., 51, 52
Jansson, E., 31*n*5, 140, 157
JND. *See* Difference limen
Just discriminable change, 121
Just noticeable difference. *See* Difference limen

Kaas, J., 95–96
Kahn, J., 161
Kajiyama, M., 53*n*9
Kakusho, O., 121, 122
Kato, K., 121, 122
Kay, N., 5*n*5
Keidel, W., 106–107, 125
Kiang, N., 99
Kimura, D., 150
Klein, W., 135, 158
Kobayashi, T., 121, 122
Kruskal, J., 131
Kuhl, P., 151–52
Kuzirai, K., 107

Ladefoged, P., 88*n*39, 133
Lane, H., 147–48
Langner, G., 106–107, 112, 114
Lansky, P., *Six Fantasies* . . . , 188*n*7
Lass, R., 56
Lateral inhibition, 123–24
Laterality. *See* Hemispheric specialization of cortex
Lauter, J., 151
Laws for tone color, 4
LAX sounds in Babbitt's *Ensembles* . . . , 180
LAXNESS: as a vowel feature, 52; contours of, 56; and vocal relaxation, 56; transposition, 71–72, 161, 193–194; and inversion in other dimensions, 78; evidence for, 136, 158; and the speech mode, 158; and closure, 197; and "stretching," 198–99
Learning: of F-patterns, 106; problem of, 119–20; and categorical perception, 148
Left-ear advantage, 150
Legitimacy of operations, 80
Lehiste, I., 148
Length of tube. *See* Tube length
Lenneberg, E., 61*n*29, 146*n*1
Lewin, D., 76
Liberman, A., 87–88, 145, 147
Liberman, M., 95
Lichte, W., 134
Licklider, J., 130
Limitations: of Rule 1, 42–43; of the dimensions, 65–67; of physiological methods, 91–93; of psychoacoustic methods, 116–20; of speech and cognitive science, 145–52; of music-analytic evidence, 166–68; of a composition as evidence, 191
Limited variation, and "relative-pitch" timbre, 187–88
Lindsey, P., 49*n*16, 50*n*17, 146
Linking, 170–73
Linschoten, D., 104–105
Liquids, and sound-color dynamics, 87–88
Locker cells, 110
Locking, in periodicity detection, 113–14
Lovendusky, J., 206–207
Low-pass filters, 35*n*8
Lublinskaya, V., 125
Luce, D., 117, 118

Magnitude estimation: method of, 118; of differences in timbre and vowel quality, 130–31

Magnitude production, method of, 118
Manley, J., 100
Marks, L., 20, 117, 118–19
Marler, P., 152n3
Marmor, G., 163
Mass in sound objects, 6–7
Mathews, M., 229
Matrices of series for *Colors*, 211–12
Maximal invariance in four-color sets, 207
MDS. *See* Multidimensional scaling
MDSCAL, 131
Measurement in sound-color space, 58–59
Medial geniculate, 95–96, 106
Melodies: color, 3, 215; dichotic, and hemispheric specialization, 150; differentiation of, 169
Mel scale, 129
Membranes, dimensions in, 66–67
Merzenich, M., 95–96, 99, 107n6
Metathetic continua, 58n28
Methods: physiological, psychoacoustic. *See* Limitations
Metric of sound-color space, 58
Metzler, J., 163
Miller, J. D., 151–52
Miller, J. R., 138
Miller, M. I., 102n4
Miller, R. L., 156
Mirror images, 161
Missing fundamental, case of, 109–111
Mixture: of visual colors, 48–49; of sound colors, 88–89
Modulo-12 addition, 72–73
Møller, A., 99, 110
Moorer, J., 138
Morgan, M., 161
Morley, T., 14n8
Morris, R., 202n9, 213
Morse code, and brain laterality, 150
Motor theory. *See* Active theories of speech perception
Mouth opening, and OPENNESS, 52
Mueller-Preuss, P., 100
Multidimensionality: of visual color, 48–49; of taste, 49; of vowels, 50–51
Multidimensional scaling: of timbre, 134–36, 138; of speech, 158–59
Multiple operations, 199–207
Multiple resonances, 35–38; in digital synthesis, 229
Musical instruments. *See* Instruments
Musical object, 7–8
Musical space, 12, 13
Music-11, 229
Musique concrète, and Schaeffer's *Traité*, 5–6
Mynah bird, cortical responses in, 106–107

Naming convention for operations, 200
Nasal sounds, 42–43
Naturalness of sound-color operations, 80–81
Negation, and inversion, 76
Negative results, status of, 92
Nettleton, N., 151
Neutrality of stimuli. *See* Stimuli: ethological neutrality of
Neutral point, 58; in shifted space, 79
Newman, E., 127–28
Nielson, D., 95n2
Nine-color collection, 193–94, 209–210

Noise, experimental, 60
Noise source, 46; in *Colors*, 208–209, 220. *See also* Source: adequacy of
Non-coloristic timbre, 170
Non-linearity in auditory system, 100
Non-uniform tubes, 39
Normal collections, 192–99; and stretching, 199; combinations of operations in, 203
Normalization, 62–65, 66; of vowels, 64–65, 155; effects on operations, 79; cues for, 133, 155–56
Norman, D., 49n16, 50n17, 146
Notation for operations, 191–92
Nuages (Debussy), analysis of, 12

Objet musical, 7–8
Objet sonore, 6–7
Oboe, resonances in, 157
Octave: and color space, 69; and color wrap-around, 72–73
Off-center axes of inversion, 82–83
Off-peak partials, 104
Olivary nuclei, 95
Oliveros, P., 185
Olson, R., 128
Onto mapping, 82–84
OPENNESS: choice of terminology, 52; contours of, 54–55; transposition, 70–71; inversion, 78; evidence for, 135–36, 158; independent musical variation in, 176–77
Operations, 17–18, 68–86; combinations of, 84, 163, 201–203, 206; psychoacoustics of, 140–41; cognitive analogies to, 159–63; awareness of, 160; and musical evidence, 183–84; in resynthesized speech, 188; cyclic definition of, 194–95; naming convention, 200; on color series in *Colors*, 210–12; and poetry, 223
Opposition in sound-color direction, 173–74
Orchestral brightness, 172
Ordered series of *Colors*, 210–16
Orthogonality, and dimensions, 59–60, 70–71
Outer squares in sound-color collections, 197
O'Neill, W., 107

Papcun, G., 150
Parallel filter networks, 227
Parallel tracks in auditory pathways, 95–96
Particle displacement in uniform closed tube, 36
Pathways of auditory nervous system, 94–97
Peak amplitude. *See* Resonance: amplitude of
Perceptibility: and wrap-around, 76; and the operations, 80
Perceptual mode, and context, 119
Perceptual space. *See* Operations; Space
Period in damped sinusoid, 45
Periodicity detection, 110, 113–14; and residue pitch, 110–11; psychoacoustics of, 129–30
Peterson, G., 53n25, 136, 153–54, 156
Phase at resonance, 34
Physiology, and psychological claims, 92
Pick, G., 122n5
Pisoni, D., 147

Pitch: change of and source/filter distinctions, 30, 47–48; transposition, 69–70; inversion, 76; physiology of, 109–111; in ringing filters, 184; in *Colors*, 208–209, 219–22; in pulsed sources, 221
Pitch class, and sound color, 69, 192*n*2
Plate resonance: in violin, 47; and the dimensions, 66–67
Plomp, R., 131–32, 133, 135, 158
Podgorny, P., 163
Poetry, and sound color, 222–23
Pols, L., 135, 158
Positive results, and psychological claims, 92
Practice, and brain laterality, 150
Precompositional design, in *Colors*, 216
Predisposition for language, 61. *See also* Innate releasers
Pre-speech, and sound-color dimensions, 60–62
Primitive attribute, sound color as a, 62
"Primitive" processes, 93
Production, effect of on perception. *See* Active theories of speech perception
Prothetic continua, and LAXNESS, 58*n*28
Psychoacoustics, 116–43; in Schaeffer's *Traité*, 9; and definition of elements, 15–16. *See also* Limitations
Psychological dimensions. *See* Dimensions
Psychological distance, measures of, 118–19; along dimensions, 142–43

Psychological theory, and music theory, 84*n*37
Pulses: adequacy as sources, 45–46; as psychoacoustic stimuli, 127*n*7; in Subotnick's *Until Spring*, 173; in *Colors*, 221

Q. *See* Quality factor
Quality, instrumental, 31
Quality factor, of a resonator, 32
Quarter-wavelength tube, 36
Questions, principal, of the theory, 20–21; physiological, 91, 113–15; psychoacoustical, 142–43

Rabiner, L., 226
Rameau, J., 14*n*8
Randall, J. 20*n*11
Range, stimulus, and MDS studies, 159
Rate as neural response measure, 101
Rationalist tradition, 8–9
RCA synthesizer, 169–70
Realization. *See* Setting
Reduced hearing, 6; and the mel scale, 129
Relative pitch theory, 47–48; and dissimilarity judgments, 132; and computer music, 187–88
Residue, 110, 129–30
Resolution of resonances, 65–66, 104
Resonance: and spectrum envelope, 31–48; frequency of, 32; amplitude of, 33, 37–38; analysis of in auditory system, 100–106; resolution of, 104; and cochlear nucleus response, 105; psychoacoustic mechanisms for detection of, 123–25
Resonance curves, 32

Resonators: simple, 32–35; adding of multiple, 36–38; aberrant, and the dimensions, 66–67
Resynthesis of musical instrument sounds, 138
Retroflexion as a dimension of vowels, 158
Rhythmic motives in *Colors*, 219–20
Ringing, in filters, 184
Risset, J.-C., 139
Ritsma, R., 130
Rock, I., 161–62
Rose, J., 99
Rosen, G., 38n11
Rosner, B., 148
Rotated images in vision and touch, 161–62
Rotation: of mental images, 162–63; sound-color, 163
Roughness, 134
Rounding, in vowels, 51
Rubber band/hammer model of resonance, 34
Rule 1, 41–42; and auditory nerve response, 102; as primitive auditory feature, 103; psychoacoustics of, 131; and Babbitt's *Composition for Synthesizer*, 170; and changing sound colors, 174–75
Rule 1', 23
Rule 1", 26; and auditory nerve response, 102
Rule 2, 57; limitations of, 65–67
Rule 3a, 75; in shifted spaces, 79; as obligatory or permissive, 160
Rule 3b, 76–77; in shifted spaces, 79; as obligatory or permissive, 160
Rule X as an alternative, 29–30
Rumelhart, D., 140

Sachs, M., 101–103, 113, 125, 152
Saldanha, E., 139
Salience: and the dimensions, 60, 159, 224; and MDS studies, 138
Sampling rate, and digital filters, 229
Sanders, D., 146
Saslow, M., 121, 122
Saturation of neurons and spread of excitation, 98
Scales: psychoacoustic, for resonance frequency, 125–27; multidimensional, for timbre, 134–38
Schaeffer, P., 5–10
Scharf, B., 121
Scheich, H., 106–107, 112, 114
Schenker, H., 85
Schoenberg, A., 3, 68, 213
Schouten, J., 110, 129–30
Score of *Colors*, 217–18
Searle, C., 113n8
Seebeck, A., 129
Selective advantage of F-pattern discrimination, 106
Selective areas in mynah bird cortex, 108, 112
Semantic differential, 136–37
Semantics in poetry, and sound color, 223
Sensitivity: to changes in filter parameters, 120–23; interpretation of measures, 122–23; in LAXNESS-closed collections, 198n5
Sequencer as filter control, 228
Serialism, and intuition, 191
Set, perceptual, 172
Setting, of color structure in *Colors*, 216–22
Shankweiler, D., 149

"Sharpening," 123–24
Sharpness, 126; in semantic differential analysis, 136
Shepard, R., 16, 117, 119, 136, 162–63
Shifting of color space, 63–65, 66; operations and, 79; and color-space hierarchies, 86–87; and the fundamental, 133
Short vowels, 51. *See also* Inner squares in sound-color collections
Shower, E., 121n4
Signal processors, implementing filters in, 229–30
Signal-to-noise ratio in analog filter networks, 227
Sign convention for dimensions, 58–59
Similarity: judgments in multidimensional space, 125–26; and SMALLNESS operations, 202
Singh, S., 158
Singing, 40–41
Single dimensions, changes in, 176
Sinnott, J., 152
Sinusoid: and source adequacy, 44; sound color of, 171
Size of discriminable sound-color collections, 196
Slawson, A. W., 10, 36n10, 55, 60, 129, 130–31, 133, 156, 160, 191, 229
SMALLNESS: contours of, 56; and normalization, 65; transposition in, 73–74, 195, 197, 200n8; inversion in, 78; MDS evidence for, 135; and sharpness, 137; independent musical variation in, 176–77; and stretching, 224
Smith, J. C., 111

Smith, J. O., 228
Smoorenberg, G., 104–105
Solfège of musical objects, 8
Solipsism in music-analytic evidence, 167
Somatosensory images, and visual images, 161–62
Sound mass in Babbitt's *Ensembles*, 178–79
Sound object, 6–7
Sound spectrographs: in vowel studies, 153; interpretation of, 167–68
Source: /filter model, 22–31, 85; in musical instruments, 27–31, 181; adequacy of, 43–46; ideal, 44–46; distinguishing from filter, 46–48; knowledge of, 47; and sound-color mixture, 88–89; analyzers, 109; psychoacoustics of, 127–30, 130–33; musical contrast in, 170–71; in speech resynthesis, 188
Source function, in difference equations, 228–29
Space: color, 73; operations in perceptual, 69, 77; color, and vowels, 107–108; color, size of, 154; musical evidence for perceptual, 175; color, musical filling, 178–80
Space limitation, and hierarchical structuring, 86–87
Species, of a musical object, 8
Species-specificity: of feature detectors, 100; and vowel perception, 152–53
Spectrum: and Cogan and Escot's "color," 12–13; and dissimilarity judgments, 135

Spectrum envelope: of filters, 25–26; constancy of, under pitch change, 27; of multiple resonators, 36–38; effect of F-pattern on, 40–41; representation in auditory nerve, 102
Speech, resynthesis of, 188
Speech mode: as predisposition for language, 61; as a "set" for speech behavior, 145; evidence for, 146–52
Speech origins, 62
Speech perceptions in animals, 151–52
Speech science, and sound color, 144
Speed change of tape recorders, and color, 172
Split formants, 121
Spread of excitation in cochlear nerve and cochlear nucleus, 97–98
Square sound-color collections, normality in, 196
Starr, A., 99
Starr, D., 213
Steady-state colors in *Colors*, 220–21
Steeneken, H., 131–32
Stern, R., 113n8
Stevens, K. N., 35n9, 38, 61–62, 120, 122, 145
Stevens, S. S., 58n28, 60, 109, 117, 118–19, 127–28
Stimuli: choice of, 100; ethological neutrality of, 103
Stockhausen, K.: *Hymnen*, 172; *Kontakte*, 184; *Stimmung*, 187; *Telemusik*, 180–83
Stravinsky, I., 16n10
Stretching: and sound-color operations, 142; and LAXNESS transposition, 198–99; and SMALLNESS, 224
Strong, W., 31n5
Strong-coupled systems, 47
Structural link, sound color as, 170–73
Studdert-Kennedy, M., 149, 164
Stumpf, C., 35n9, 129, 172
Subotnick, M.: *Until Spring*, 173–74; *A Sky of Cloudless Sulphur*, 176–77
Subsets: and closure, 192–93; of color collections, 206–207
Subspaces, and color hierarchies, 87
Substratum, sound color as a, 62
Successive filtering, and color hierarchies, 85–86
Suga, N., 107
Summerfield, A., 156
Sundberg, J., 140
Suppression: in cochlear nucleus, 98; of off-formant partials, 102; pattern of in mynah bird cortex, 107
Swapping in combinatorial matrices, 213–14
Swedish vowel system, 153–54, 196
Sweet Honey in the Rock, 185
Swets, J., 117
Swift, R., 10, 11
Symmes, D., 164
Symmetry: and normal sound-color collections, 195; visual, and sound-color inversion, 161–62
Synesthesia, and sound color, 20
Synthesizers, 173, 226–28
Synthetic sounds, and scaling of sound color, 126
Synthetic vowels: as physiological stimuli, 101–103; frequency following response to, 111

Template, filter as, 168
Temporal contrasts, and sound-color mixture, 89
Temporal dimension, and color in hearing and vision, 20
Temporal envelopes in MDS studies, 138
Temporal order, and vowel perception, 156–57
Tenney, J., 5n4
Terhardt, E., 111
Tetra-colors, 206–207
Text in vocal music, 185
Text-sound composition, 223
Theme, combinatorial structure as, 214
Theory: formal requirements of, 14–18; domain of, 18–20; compositional vs. psychological, 84n37
Third formant, and normalization, 156. *See also* Higher resonances
Thomas, I., 156
Thresholds, 117–18. *See also* Difference limen; Sensitivity
Timbre: and sound color, 18–20; meanings of, 19; of residue, 129–30; in psychological studies, 131n8
Timbre analogies, method of, 140
Timing, and hierarchical filtering, 85–86
Tone color: melody, 3; dimension of auditory sensation or instrumental combination, 4n3
Tone languages, 196n4
Tone quality, and timbre, 19
Tonotopic organization, 94
Transient detectors, 106
Transition in consonants, 88

Transposition: sound-color, 69–76; reality of sound-color, 79–80; of pitch and sound color, 128, 160; and timbre analogies method, 141; cyclic definition of, 194–95; combinations of, 201
Treatment in sound objects, 6–7
Tromp, H., 158
Tube length: and SMALLNESS, 56; and formant frequencies, 63
Twelve-color collection, 196–97
Twelve-tone aggregates, 213
Two-dimensional color space, 179–80, 182–83
Two-formant filters, 208
Two-formant vowels, and "center of gravity" effect, 125
Two-tone stimuli, 98
Type, of a musical object, 8

Uniform tube: resonances in, 35–36; and maximally LAX point, 56
Unique inversions in nine-color collection, 203–204
Unique transpositions in nine-color collection, 204–205
Unit generators for resonance filters, 229
Units, elementary, in sound-color theory, 15

Validity of psychological scales, 118–19
Variation: basis in principle of invariance, 18; independent, vs. independent constancy, 59–60; and musical operations, 68, 70–71
Variations, character of in *Colors*, 208–209

Verbrugge, R., 153*n*4
Vercoe, B., 229
Verification, and analysis, 167
Verticality, and perception of visual symmetry, 162
Vibrators, on skin as a cochlear model, 123–24
Violin, fixed resonances in, 157
Visual color, and sound color, 20, 48–49
Vocal chords, 24
Vocal filter: and spectrum envelope, 25–26; used as selector of harmonics, 187
Vocal imitation of musical instruments, 67
Vocality, and timbre, 19
Vocalizations of animals, 100
Vocal music, and sound-color theory, 166, 184–86
Vocal source, 24–27; and bow/string system in violin, 157. *See also* Source
Vocal tract, size of, 64–65
Voigt, H., 102
Volkmann, J., 127–28
Voltage-controlled filters, 173, 226–28
Vowel diagram, 153, 154
Vowel perception, 153–55
Vowels: as models for sound color, 24–27, 154; production of, 28; auditory nerve responses to, 101–103; detectors for, 106; studies using the semantic differential method, 136; and the speech mode, 148, 149–51; and dichotic listening, 149, 151; animal discrimination of, 151–52; American and Swedish, 153–54; perception of, 153–55

Waveform: of impulse-excited resonator, 45; and fixed vs. relative pitch transformations, 47–49
Weakly coupled systems, 42–43. *See also* Coupling
Webber, L., 185*n*6
Wessel, D., 139, 140
Whisper as a source, 46
Whispered vowels, representation of in auditory nerve, 102
Whitfield, I., 95, 97–98, 109, 110
Wiesal, T., 61*n*30, 99–100
Woods, D., 158
Wrap-around: in color transposition, 72–75; for alternate inversion axes, 82–83
Wrap-inside-out in LAXNESS transposition, 75
Wuorinen, C., *Time's Encomium*, 170

Xenakis, I., 7

Yilmaz, H., 88*n*40
Yost, W., 95*n*2
Young, E., 101–103, 113, 125

Zaback, L., 163
Zeroth contour as reference for dimensions, 58–59
Zero-width pulse, 44–45
Zwicker, E., 121, 126

781.15 S631s
Slawson, Wayne.
Sound color

Designer:	Wilsted & Taylor
Composition:	Wilsted & Taylor
Typeface:	12/15 Mergenthaler Imprint
Printer:	Malloy Lithographing
Bindery:	John H. Dekker & Sons
Text paper:	50 lb glatfelter, opacified shade A50
Soundsheets:	Eva-Tone